CARAWAY

CARAWAY
The Genus *Carum*

Edited by

Éva Németh
*University of Horticulture and Food Industry
Budapest, Hungary*

CRC PRESS

Boca Raton   London   New York   Washington, D.C.

FIRST INDIAN REPRINT, 2012

This book contains information obtained from authentic and highly regarded sources. Reprinted material is quoted with permission, and sources are indicated. A wide variety of references are listed. Reasonable efforts have been made to publish reliable data and information, but the author and the publisher cannot assume responsibility for the validity of all materials or for the consequences of their use.

Neither this book nor any part may be reproduced or transmitted in any form or by any means, electronic or mechanical, including photocopying, microfilming, and recording, or by any information storage or retrieval system, without prior permission in writing from the publisher.

Direct all inquiries to CRC Press LLC, 2000 N.W. Corporate Blvd., Boca Raton, Florida 33431.

© 1998 CRC Press, LLC

Trademark Notice: Product or corporate names may be trademarks or registered trademarks, and are used only for identification and explanation, without intent to infringe.

Visit the CRC Press Web site at www.crcpress.com

Printed and bound in India by
Replika Press Pvt. Ltd.

ISBN 10 : 90-5702-395-4
ISBN 13 : 978-90-5702-395-8

FOR SALE IN SOUTH ASIA ONLY.

CONTENTS

PREFACE TO THE SERIES

There is increasing interest in industry, academia and the health sciences in medicinal and aromatic plants. In passing from plant production to the eventual product used by the public, many sciences are involved. This series brings together information which is currently scattered through an ever increasing number of journals. Each volume gives an in-depth look at one plant genus, about which an area specialist has assembled information ranging from the production of the plant to market trends and quality control.

Many industries are involved such as forestry, agriculture, chemical, food, flavour, beverage, pharmaceutical, cosmetic and fragrance. The plant raw materials are roots, rhizomes, bulbs, leaves, stems, barks, wood, flowers, fruits and seeds. These yield gums, resins, essential (volatile) oils, fixed oils, waxes, juices, extracts and spices for medicinal and aromatic purposes. All these commodities are traded world-wide. A dealer's market report for an item may say "Drought in the country of origin has forced up prices".

Natural products do not mean safe products and account of this has to be taken by the above industries, which are subject to regulation. For example, a number of plants which are approved for use in medicine must not be used in cosmetic products.

The assessment of safe to use starts with the harvested plant materials which has to comply with an official monograph. This may require absence of, or prescribed limits of, radioactive materials, heavy metals, aflatoxins, pesticide residue, as well as the required level of active principle. This analytical control is costly and tends to exclude small batches of plant materials. Large scale contracted mechanised cultivation with designated seed or plantlets is now preferable.

Today, plant selection is not only for the yield of active principle, but for the plant's ability to overcome disease, climatic stress and the hazards caused by mankind. Such methods as *in vitro* fertilisation, meristem cultures and somatic embryogenesis are used. The transfer of sections of DNA is giving rise to controversy in the case of some end-uses of the plant material.

Some suppliers of plant raw materials are now able to certify that they are supplying organically-farmed medicinal plants, herbs and spices. The Economic Union directive (CVO/EU No 2092/91) details the specifications for the **obligatory** quality controls to be carried out at all stages of production and processing of organic products.

Fascinating plant folklore and ethnopharmacology leads to medicinal potential. Examples are the muscle relaxants based on the arrow poison, curare, from species of *Chondrodendron*, and the antimalarials derived from species of *Cinchona* and *Artemisia*. The methods of detection of pharmacological activity have become increasingly reliable and specific, frequently involving enzymes in bioassays and avoiding the use of laboratory animals. By using bioassay linked fractionation of crude plant juices or extracts, compounds can be specifically targeted which, for example, inhibit blood platelet aggregation, or have antitumour, or antiviral, or any other required activity. With the assistance of robotic devices, all the members of a genus may be readily screened. However, the plant material must be **fully** authenticated by a specialist.

The medicinal traditions of ancient civilisations such as those of China and India have a large armamentarium of plants in their pharmacopoeias which are used throughout South East Asia. A similar situation exists in Africa and South America. Thus, a very high percentage of the world's population relies on medicinal and aromatic plants for their medicine. Western medicine is also responding. Already in Germany all medical practitioners have to pass an examination in phytotherapy before being allowed to practise. It is noticeable that throughout Europe and the USA, medical, pharmacy and health related schools are increasingly offering training in phytotherapy.

Multinational pharmaceutical companies have become less enamoured of the single compound magic bullet cure. The high costs of such ventures and the endless competition from me too compounds from rival companies often discourage the attempt. Independent phytomedicine companies have been very strong in Germany. However, by the end of 1995, eleven (almost all) had been acquired by the multinational pharmaceutical firms, acknowledging the lay public's growing demand for phytomedicines in the Western World.

The business of dietary supplements in the Western World has expanded from the Health Store to the pharmacy. Alternative medicine includes plant based products. Appropriate measures to ensure the quality, safety and efficacy of these either already exist or are being answered by greater legislative control by such bodies as the Food and Drug Administration of the USA and the recently created European Agency for the Evaluation of Medicinal Products, based in London.

In the USA, the Dietary Supplement and Health Education Act of 1994 recognised the class of phytotherapeutic agents derived from medicinal and aromatic plants. Furthermore, under public pressure, the US Congress set up an Office of Alternative Medicine and this office in 1994 assisted the filing of several Investigational New Drug (IND) applications, required for clinical trials of some Chinese herbal preparations. The significance of these applications was that each Chinese preparation involved several plants and yet was handled as a **single** IND. A demonstration of the contribution to efficacy, of **each** ingredient of **each** plant, was not required. This was a major step forward towards more sensible regulations in regard to phytomedicines.

My thanks are due to the staff of Harwood Academic Publishers who have made this series possible and especially to the volume editors and their chapter contributors for the authoritative information.

Roland Hardman

CONTRIBUTORS

Harro J. Bouwmeester
Institute for Agrobiology and Soil
  Fertility (AB-DLO)
P.O. Box 14
6700 AA Wageningen
The Netherlands

Michael Dachler
Bundesamt und Forschungszentrum
  für Landwirtschaft
Spargelfeldstrasse 191,
1220 Wien
Austria

Leon G.M. Gorris
Agrotechnological Research Institute
  (ATO-DLO)
P.O. Box 17
6700 AA Wageningen
The Netherlands

Seija Hälvä
Laboratories for Natural Products, and
  Aromatic and Medicinal Plants
University of Massachusetts
Amherst, MA
USA

Klaasje J. Hartmans
Agrotechnological Research Institute
  (ATO-DLO)
P.O. Box 17
6700 AA Wageningen
The Netherlands

J. Henk Lubberts
Centre for Plant Breeding and
  Reproduction Research (CPRO-DLO)
P.O. Box 16
6700 AA Wageningen
The Netherlands

Erzsébet Mihalik
Department of Botany and Botanic
  Garden
József Attila University
P.O. Box 657
6701 Szeged
Hungary

Éva Németh
Department of Medicinal Plant
  Production
University of Horticulture and
  Food Industry
Villányi str. 29-43
H-1114 Budapest
Hungary

Grazyna Obidoska
Department of Genetics, Plant Breeding
  and Biotechnology
Warsaw Agricultural University
Nowoursynowska 166
02-766 Warsaw
Poland

Koos Oosterhaven
Agrotechnological Research Institute
  (ATO-DLO)
P.O. Box 17
6700 AA Wageningen
The Netherlands

Eli Putievsky
Division of Aromatic Plants
Agricultural Research Organization
Newe Ya'ar Research Center
P.O. Box 1021
Ramat Yishay 30095
Israel

Joanna Ruszkowska
The Faculty of Chemistry
University of Warsaw
ul. Pasteura 1
02-093 Warsaw
Poland

Ala Sadowska
Department of Genetics, Plant Breeding
  and Biotechnology
Warsaw Agricultural University
Nowoursynowska 166
02-766 Warsaw
Poland

Eddy J. Smid
Agrotechnological Research Institute
  (ATO-DLO)
P.O. Box 17
6700 AA Wageningen
The Netherlands

Hille Toxopeus
Nassauweg 14
6703 CH Wageningen
The Netherlands

Zenon Wêglarz
Department of Medicinal Plants
Warsaw Agricultural University
Nowoursynowska 166
02-787 Warsaw
Poland

1. INTRODUCTION

ÉVA NÉMETH

*University of Horticulture and Food Industry, H-1114 Budapest,
Villányi str. 29–43, Hungary*

Although the genus *Carum* counts 25 species, through the Northern hemisphere (Danert *et al.* 1975), only *Carum carvi* L. has an economical importance, being used and cultivated in several regions (Figure 1). This species has been the topic of a wide row of investigations, and – in the course of production – of accumulated practical knowledge. The majority of these valuable data is collected in this book, part of them published elsewhere, part of them nowhere.

Carum carvi L., caraway is believed to have been cultivated and consumed in Europe longer, than any other spice species. Seeds, found in ancient debris in Switzerland should be a proof for it (Rosengarten 1969). Cultivation is known since the Middle Ages, from Sicily to northern Scandinavia. It is mentioned even in a XIV. century English cookery book and also in classical literature, in Shakespeare's Henry IV. (Rosengarten 1969).

Caraway is widely used as a condiment, as a drug, and recently, also for some other industrial purposes. Separate chapters of this book deal with each utilisation area. Traditionally it is grown mainly in the Netherlands, – which has an outstanding reputation – Germany, Poland, Ukraine, Hungary, Rumania. Heeger (1956) mentions also some further countries, such as Sweden, Norway, Spain and Austria, however their production seems not to be a determining factor today on the world market. In the contrary, the production is growing in some other regions, such as Canada, United States, Finland, Syria, Morocco (Parry 1969). It is likely, that in these last two countries the annual variety is cultivated, although this fact is not mentioned in the cited manuals (Figure 2). In general, information on the presence of annual and biennial eco-types within the species are few, and the existing ones rather new, originating from the last 20–25 years. Recently, production of the annual type is spreading in the Middle and West European countries, because of the simpler agriculture and higher yields. Origin of the annual material should be Egypt, which is also a main producer of it. However, production was decreasing in the period of 1975–91 (Abu-Nahoul and Ismail 1995).

In the countries of the near East caraway is often substituted by cumin (*Cuminum ciminum* L.) this small, annual species, (Figure 3) which is indigenous in Egypt, widely grown in the Mediterranean, in Asia and also in America. The fruits and occasionally the whole herb are used as a spice and aperitif in food industry, but also as a carminativum and stomachicum in phytotherapy (Pillai and Nambiar 1982). Although usage and aroma is similar to those of caraway, main component of the essential oil is cumin-aldehyde (Parry 1969).

Figure 1   *Carum carvi* L. f. *biennis*

Caraway is called also carvi (in French and Italian), Kümmel (German), alcaravea (Spanish), (Tucker 1988), karvij (Dutch), kminek (Polish), kömény (Hungarian) etc. All European countries have their own, – however to some extent similar – words for this species, which names might be drawn back to the Arabian "karauya" from the XII century (Rosengarten 1969). Turnover of caraway products, seeds and essential oil are not unimportant ones among spices and medicinal plants. According to Lawrence (1993) around 30 tonnes of essential oil are traded yearly in the world, with this amount caraway stands on the fifth place among *Apiaceae* species. The world production of seeds may be assumed reaching 10–15 thousand tonnes. The production however, is rather varying, fluctuations from year to year both in quantities and in prices are characteristic.

Figure 2   *Carum carvi* L. f. *annua*

Basic quality factor of the seeds as drugs is the essential oil content. Its required minimum values show some differences from country to country. In Deutsches and in Austrian Arzneibuch a limit of 4.0% d. w. is required, in Ph. Helv. (Swiss Pharmacopoeia) 3,5% (Wichtl 1989), in British Pharmacopoeia 3,5% (Evans 1996), in Ph.Hg. in Hungary 2.5%. Annual varieties usually are not able to meet these values. The problem could be illustrated by the Hungarian case, where two different standards refer to the biennial and the annual forms. However, a better solution would be a harmonised quality specification among the different countries, e.g. as an ESCOP monograph. A monograph on *Carvi fructus* has been under elaboration in the last period (Krant 1996).

A further quality factor is the carvon proportion of the essential oil, which is usually fixed in at least 50%, where again the annual varieties are sometimes in the trouble not to reach it.

Figure 3 *Cuminum cyminum* L.

Improvement of quality parameters, such as increase of essential oil content up to 5 ml/100 g and its carvon percentage above 60% is recently one of the main topics of breeding work and cultivar development (*Anonymous* 1996). Besides, new cultivars should assure high and stable yields up to 2 tonnes/ha, which is connected with a non-shattering characteristics. Remaining the carpophores on the seeds is a drawback and annual cultivars are more likely to possess this feature. Also resistance against some pests and insects (*Depressaria nervosa, Erysiphe umbelliferarum, Eryophyes peucedani f. carvi, Mycocentrospora acerina, Sclerotinia sclerotiorum*, etc.) raises the production value of a cultivar (Sváb 1993, Evenhuis *et al.* 1995). However, till now this problem was not among the most important ones and no special material selected for this aim is known. Four regis-trated varieties of *Carum carvi* L. are known in the European variety list, 10 ones accord-ing to the OECD variety list, 15 ones in the German "Sortenlist" (*Anonymous* 1996). In some of these lists there are overlappings, such as altogether 3 Dutch, 3 Czech,

Table 1 Cultivars of caraway

Name	Form	Origin	Source of information
Bleija	biennial	The Netherlands	EU list, OECD list, Sortenliste
Kami	biennial	Denmark	EU list, OECD list, Sortenliste
Sylvia	biennial	Denmark	EU list, OECD list, Sortenliste
Volhouden	biennial	The Netherlands	EU list, OECD list
Hollandi	biennial	Hungary	OECD list, Sortenliste
Maud	biennial	Hungary	OECD list
Konczewicki	biennial	Poland	OECD list
Plewiski	biennial	Poland	OECD list
Mausholt's Karwijsaad	biennial	Denmark	OECD list
Niederdeutscher	biennial	Germany	Sortenliste
Rekord	biennial	Czech Rep.	Sortenliste
Kepron	biennial	Czech Rep.	Sortenliste
Prochan	biennial	Czech Rep.	Sortenliste
De Ghimbav	biennial	Rumania	Sortenliste
Mare de Roman	biennial	Rumania	Sortenliste
Polaris	biennial	Norway	Literature
Podolskii 9	biennial	Ukraine	Literature
Khmelnitskii	biennial	Ukraine	Literature
SZK-1	annual	Hungary	Sortenliste
Karzo	annual	The Netherlands	OECD list
CN-1	annual	Israel	Literature
Bi-An	annual	Israel	Literature
Balady	annual	Egypt	Literature

3 Hungarian, 2 Polish, 2 Rumanian, 1 German and 3 Danish varieties, – among them only 2 annual ones – are in the trade (Table 1). Furthermore, there is some information about Israeli, Norwegian and Egyptian cultivars or candidates on the base of scientific publications (El-Ballal 1980, Franz 1996, Kallio *et al.* 1994, Kuzmich 1986, Putievsky *et al.* 1994). Caraway had been and remained a condiment and a medicinal drug of high importance. Its usefulness and enjoyable aroma assures also a firm future for this species in food industry, in human and veterinarian therapy and in agricultural production.

This book was edited in order to collect all the existing scientific material on *Carum* genus and especially on *Carum carvi* L. We intended to discuss all the important and interesting topics in connection with biology, production and utilisation. The book should be of benefit for all specialist and people who are interested in up-to-date information.

REFERENCES

Abu-Nahoul, M.A. and Ismail, T.H. (1995) The features of foreign trade for some aromatic and medicinal plants in Egypt. *Assiut Journal of Agricult. Sciences*, **26**, 319–335.

Anonymous (1996) *Beschreibende Sortenliste Heil – und Gewürzpflanzen*, Landbuch Verlag, Hannover, pp. 66–70.

Danert, S., Hanelt, P., Helm, J., Kruse, J. and Schultze-Motel, J. (1975) *Urania world of plants, Vol. 2. Higher plants (Uránia növényvilág, Magasabbrendű növények).* Gondolat Publisher, Budapest, p. 162.

El-Ballal, A.S.I. (1980) Genotype–environmental interaction in essential oil yield in the selected caraway type, 1.4.36, *Herba Hungarica*, **18**, 155–166.

Evans, W.Ch. (1996) *Trease and Evans Pharmacognosy*, WB Saunders Co. Ltd., London, Philadelphia, Toronto, Sydney, Tokyo, pp. 263–267.

Evenhuis, A., Verdam, B., Gerlagh, M. and Goossen-van de Geijn, H.M. (1995) Studies on major diseases of caraway (*Carum carvi*) in the Netherlands, *Industrial Crops and Products*, **4**, 53–61.

Franz, Ch. (1996) Züchtungsforschung und Züchtung an Arznei- und Gewürzpflanzen in ausgewählten Ländern Europas und des Mittelmeergebietes, *Arznei- und Gewürzpflanzen*, **1**, 30–38.

Heeger, E.F. (1956) *Handbuch des Arznei – und Gewürzpflanzenbaues*, Deutscher Bauernverlag, Berlin, pp. 328–338.

Kallio, H., Kerrola, K. and Alhonmaki, P. (1994) Carvone and limonene in caraway fruits (*Carum carvi* L.) analysed by supercritical carbon dioxide extraction-gas chromatography, *J. Agricult. and Food Chemistry*, **42**, 2478–2485.

Krant, W. (1996) Activities and goals of ESCOP, *ICMAP Newsletter*, **2**, 4–7.

Kuzmich, N.K. (1986) New caraway variety Podolskii 9, *Annals of the United Research Institute for Essential Oil Crops (Trudy Vsesoyuznogo Nauchno Issledovatelskogo Instituta Efiromaslicsnyh Kultur)*, **17**, 29–33.

Lawrence, B.M. (1993) A planning scheme to evaluate new aromatic plants for the flavour and fragrance industries. In J. Janick and J.E. Simon (eds.), *New crops*, John Whiley and sons inc. New York, pp. 620–628.

Parry, J.W. (1969) *Spices*, Vol. I. Food Trade Press Ltd., London, pp. 171–172.

Pharmaceutica Hungarica VII. (1986) Medicina Publisher, Budapest, **3**, 1582–1583.

Pillai, P.K.T. and Nambiar, M.C. (1982) Condiments. In Atal, C.K. and Kapur, B.M. (eds.), *Cultivation and utilization of aromatic plants*, Regional Research Laboratory, Jammu-Tawi, India, pp. 167–189.

Putievsky, E., Ravid, U., Dudai, N. and Katzir, I. (1994) A new cultivar of caraway (*Carum carvi* L.) and its essential oil. *J. Hers, Spices and Medicinal Plants*, **2**, 81–84.

Rosengarten, F. Jr. (1969) *The book of spices*, Livingston publishing Co., Wynnewood, pennsylvania, pp. 151–159.

Sváb, J. (1978) Caraway (Konyhakömény). In L. Hornok (ed.), *Cultivation and processing of medicinal plants (Gyógynövények termesztése és feldolgozása)*, Mezőgazdasági Publisher, Budapest, pp. 142–147.

Tucker, A.O. (1988) Botanical nomenclature of culinary herbs and potherbs. In L.E. Craker and J.E. Simon (eds.), *Herbs, spices and medicinal plants: Recent advances in botany, horticulture and pharmacology*, Vol. 1, Oryx press, pp. 33–80.

Wichtl, M. (1989) *Teedrogen. Ein Handbuch für die Praxis auf wissenschaftlicher Grundlage*, Wissenschafliche Verlagsges. mbH, Stuttgart, pp. 292–293.

SECTION I
BIOLOGY AND CHEMISTRY

2. TAXONOMY AND BOTANICAL DESCRIPTION OF THE GENUS *CARUM*

ERZSÉBET MIHALIK

József Attila University, Department of Botany and Botanic Garden, 6701 Szeged. P.O. Box: 657, Hungary

2.1. INTRODUCTION

Before discussing the taxonomy of the genus *Carum*, it is important to give an overview on the situation of the genus in the hierarchy of higher taxa. *Carum* genus is the member of the *Umbelliferae* family what is involved in the *Umbellales* (*Apiales, Umbelliflorae, Umbelliferales, Ammiales*) order.

The numerous name of the order shows, that the taxonomic status of the *Umbelliferae* (*Apiaceae*) family is not clear yet. The only certain point is on the level of the higher taxa, that most taxonomists agree the situation of the order, regardless the name and the families contained. The order consisting of *Umbelliferae* family is in the *Rosidae* subclass of the *Dicotyledons*.

2.1.1. Relationships at the Order Level

The affinities of the *Umbellales* are explained in different way by different authors.

According to Melchior (1964) *Umbelliflorae* (including *Cornales*) evolved from *Myrtales* and is the ancestor of *Dipsacales*.

Cronquist (1968) suggests that *Umbellales* is derived from *Sapindales* (including *Rutales*) which evolved from *Rosales*.

Takhtajan (1969) supposes direct relation between *Saxifragales* and *Cornales*. This latter involves the *Umbellales* too. According to his findings, *Cornales* and *Araliales* are separated and independent orders.

The proposal of Hutchinson (1969) describes parallel development of *Araliales* and *Umbellales* derivating both from *Magnoliales* according to the following:

Magnoliales→*Dilleniales*→*Cunoniales*→*Araliales*
↓
Ranales→*Caryophyllales*→herbaceous *Saxifragales*→*Umbellales* (only *Apiaceae*)

Hegnauer (1971) based on chemosystematic evidences, regards the *Umbellales* as the derivatives of the Rutalean stock. According to his proposal *Umbellales* is not a climax group, but a stock from which the *Asterales* evolved. There is a quite certain chemosystematic connection between *Araliales* and *Asterales*: polyacetylenes and sesquiterpene lactones which are not common in the angiosperms, frequently occur in these orders.

Cronquist (1968) maintains two distinct orders, *Cornales* and *Umbellales* (*Apiales*) and segregates them on morphological, phytochemical and anatomical basis.

2.1.2. Relationships at the Family Level

The order *Umbelliflorae* is regarded to be natural one only at limited extent because of the uncertainities in families involved. Some authors associate 7 families in this order (*Alangiaceae, Nyssaceae, Davidiaceae, Cornaceae, Garryaceae, Araliaceae* and *Umbelliferae*), while according to others, there are only two families (*Araliaceae* and *Apiaceae* or *Umbelliferae*) in it.

In his system Dahlgren (1980) puts *Araliaceae* and *Apiaceae* (Umbelliferae) in *Araliales* order, and *Cornaceae* in *Cornales*.

In the following text the name *"Umbelliferae"* will be used as a family name of the *Carum* genus.

2.2. GENERAL DESCRIPTION AND TAXONOMIC DIVISION OF THE *UMBELLIFERAE* FAMILY

2.2.1. Morphology and Anatomy of the *Umbelliferae* (Esau 1969, 1977, Tutin *et al.* 1978, Hegi 1975, Johri *et al.* 1992)

The species of the *Umbelliferae* family are usually herbs, rarely schrubs. Leaves are alternate, lamina are divided into segments and lobes, usually pinnate, much-divided. Petiole often inflated and sheathing at the base. Stipules occur only in the *Hydrocotyloideae* subfamily. Inflorescence is usually a compound umbel. Flowers are epigynous, small, hermaphrodite or unisexual. Sepals are small or absent. Petals 5, usually with three lobes. The middle lobe may inflexed. In the inflorescence the outer petals may be much more larger than the inner ones (*Orlaya*). Stamen 5, carpels usually 2, rarely 1.

In the *Umbelliferae* family there are different types of spatial separation of the stamen and ovaries. Besides monomorph flowers, when all flowers in a single plant are hermaphrodite, we can find andromonoecism (hermaphrodite and male flowers on the same plants), gynomonoecism (female and hermaphrodite flowers on the same plants) and trimonoecism (hermaphrodite, male and female flowers on the same plants). When there are different types of flowers on different plants we can separate androdioecism, gynodioecism, and trioecism: male and hermaphrodite, female and hermaphrodite flowers on different plants or in the latter case male, female and hermaphrodite flowers on separated plants in the same population. When a plant has hermaphrodite flowers, there are different mechanisms for preventing self pollination. By the *Umbelliferae* family one of these is the dichogamy. Proterandry-dichogamy and protogyny-dichogamy occur too, depending on the ripening order of anthers and stigmata.

Carpels are attached to the central axis (carpophore). The carpophore holds the mericarps after maturity of the cremocarp type of fruit. Styles usually 2, rarely 1, the base of styles often thickened (stylopodium). Ovule 1 in each loculus, apocarpous, pendent. Fruit dry. Exocarp variously indurated, endocarp woody in the subfamily Hydrocotiloideae. Mericarps joined usually by a commissure. The mericarps are more or less compressed laterally and dorsally. On the mericarps there are usually 5 ridges with 5 vascular bundles in them. Ridges are separated by valleculae. Oil ducts (vittae) are usually present in the mesocarp under the valleculae and the commissural face.

The family *Umbelliferae* is characterized by secretory tapetum in the anthers. The cytokinesis in pollen development is simultaneous. Pollen grains are two or three celled, with bilateral symmetry. Their type is tricolporate.

Ovule is anatropous, mostly unitegmic, rarely bitegmic. The embryo sac is Polygonum type. The antipodal cells are persistent only in some taxa, their secondary multiplication can also be detected. The endosperm is usually nuclear but cellular type occurs too. The embryogeny is Solanad, Onagrad or Asterad type. Testa is usually thin, seed is albuminous.

Terminal umbels are well developed, the lateral umbels are usually smaller, with fewer rays. In the smaller umbels, especially at the end of the flowering period there are numerous male flowers. Ripe fruit is essential for the certain identification of some genera.

The germination type of the *Umbelliferae* family members is epigeic. The family can be divided according to the form of the cotyledon into two groups. There are species with long, narrow cotyledons and in the other group the outline of the cotyledon is more or less round. The shape of primary leaves is characteristic for the different species.

2.2.2. Taxonomy of the *Umbelliferae*
(Borhidi 1995, Hegi 1975, Tutin *et al.* 1978)

The taxonomic-systematic structure of the *Umbelliferae* family was studied regularly. The experts of the last centuries did not divide the family into subfamiliar taxa, only groups were created on the basis of the structure of the inflorescence, the presence or absence of bracts or bracteoles and in certain extent on the form of the fruits. Later on, the anatomy of the fruit became the basic character of the subfamiliar classification. At the end of the last century Drude (1897) created his classical system mostly on this basis. The system of Drude and some others too has many uncertainities in identification, because it is not easy to delimit species and other taxa on the basis of the classical morphological data. With the development of the methods applied in taxonomy, some other characters of diagnostic value are used in classification: the embryological peculiarities, the structure and development of the pollen grains, the chemical constituents of the different species and organs, serological results etc. In spite of the fact, that this family involves numerous pharmaceutically and economically important plants, our todays knowledge is sporadic in many respect. There are numerous unsolved problems, especially the evolutionary affinities are relatively obscure yet.

The *Umbelliferae* family is regarded to be a natural one on the basis of the umbel type of inflorescence and the cremocarp fruit. This uniformity makes the subfamiliar system uncertain. Because the delimitation of the different subfamiliar taxa is difficult and therefore arbitrary, in different systems it is based on different characters. Guyot (1971) for example created a system on the basis of the stomatal types. On a single leaf in the *Umbelliferae* family several types of stomata may simultaneously be present. He divided the types based on the number of cell divisions of stomatal mother cells and on the spatial orientation of the guard cells. The occurrence of primitive, intermediary or evolved types of stomata made it possible to elucidate some phylogenetic relationships.

2.2.2.1. Chemotaxonomy of the Umbelliferae (Hegnauer 1971, Hegnauer 1964, Bohlmann 1971, Bell and Charlwood 1980, Schmitz and Seitz 1972, Klischies *et al.* 1975, Kaul and Staba 1967, Crowden *et al.* 1969)

The *Umbelliferae* family is rich in different special compounds. The most characteristic compounds are the essential oils. These compounds are secreted in schizogenous canals in all organs. There are some essential oils which has been isolated first from *Umbelliferae* species and the names of these compounds were given after these species, i.e. carvon (isolated first from *Carum carvi*). These compounds are accumulated also in other species.

The non volatile sesquiterpenes are much more characteristic to some genera in *Umbelliferae*. There are different groups of these compounds:

Daucan-type sesquiterpenes (*Daucus carota, Laserpitium latifolium*)
Ligustilide type of compounds (*Angelica, Apium, Cnidium, Levisticum species*)
Guaianolides and germacranolides (*Ferula, Laser*).

In the *Apioideae* subfamily the special coumarin compounds are the furo- and dimethylpyranocoumarines, while in the two other sub-families (*Hydrocotyloideae* and *Saniculoideae*) the furanocoumarins are accumulated. These compounds are photodinamically active. At the same time furfuranochromon improves the contraction of the heart.

The acetylenic compounds, especially the one of the chain length of C-17, give the unique chemical character of the *Umbelliferae*. In most cases these compounds result the toxicity of the species, i.e. *Cicuta* or *Oenanthe*.

The well known toxic species of the family is the *Conium maculatum* as a result of the accumulation of the piperidine alkaloid derivatives.

Saponines: These compounds are more frequent in *Hydrocotyloideae* and *Saniculoideae* than in *Apioideae*.

Fatty oils: These compounds accumulate in the seeds. The compound occuring in the greatest quantity is the petroselinic acid.

Tannins: In the *Umbelliferae* family there are no true tannins.

The chemotaxonomy is based usually on the absence or accumulation of different special compounds. Pickering and Fairbrothers (1971) use serological similarities to provide additional taxonomic data. In some cases tribes of *Umbelliferae* can be delimited well, but among *Heracleum* species they did not detected differences in using this method.

If we try to conclude these chemical findings, we must realize, that we know rather much about the chemical characters of the members of the family, but few results can be found concerning the phylogenetic or evolutionary aspects. Consequently we can rather speak about chemistry than chemosystem inside the *Umbelliferae* family. At the same time it is without doubt, that on the level of families there is close chemical similarity between *Araliaceae* and *Umbelliferae*. The existence of polyacetylenes and sesquiterpene lactones verify the close phylogenetic connection between *Araliales* and *Asterales* genera too.

2.2.2.2. Subfamilies in the Umbelliferae

The *Umbelliferae* family consist of three subfamilies: The largest one is the *Apioideae* with 250 genera, the *Hydrocotyloideae* has 34 genera and the smallest one is the *Saniculoideae* of

9 genera. The two small sub-families have no compound umbels, while in *Apioideae* this is the most striking morphological character.

According to some experts the *Hydrocotyloideae* originated from the *Araliaceae* because its fruit has woody endocarp giving intermediate between *Araliaceae* and *Umbelliferae*.

Pimenov and Leonov (1993) constructed a modern system of the family based on the work of Drude (1897). The novel item in the classification of Pimenov and Leonov (1993) is the system of the subfamilies. They describe 42 genera in *Hydrocotyloideae* with 470–490 species, the 9 *Saniculoideae* genera comprise 300–325 species and in the *Apioideae* there are more than 400 genera and about 3000 species. This description is an approximate, because it is complicated to find the limits of the species and even genera (Table 1).

2.2.2.3. Distribution of the Genera of the Umbelliferae (Pimenov and Leonov 1993)

The members of the *Umbelliferae* family are widely distributed. Asia is regarded to be the center of biodiversity of this family, but there are numerous genera and endemic taxa almost in all area of the world.

The *Hydrocotyloideae* genera occur especially in West Africa, where the other subfamilies are poorly represented (Heywood 1971).

Table 1 Tribes in the three subfamilies of the *Umbelliferae*
(Pimenov and Leonov 1993)

Hydrocotyloideae Link	*Saniculoideae* Burnett	*Apioideae* Drude
Hydrocotylinae W.D.J. Koch	*Saniculeae* W.D.J. Koch	*Echinophoreae* Benth
Mulineae DC.	*Lagoecieae* Dumort	*Scandiceae* Spreng.
		Caucalideae Spreng.
		Coriandreae W.D.J. Koch
		Smyrnieae Spreng.
		Hohenackerieae Calest.
		Pyramidopterae Boiss.
		Apieae Drude syn.:
		Ammieae, Hohenackerieae

Table 2 Geographic pattern of the genera of *Umbelliferae*
(Pimenov and Leonov 1993)

Continent/area	No. of genera	No. of endemic genera
Europe	139	29
Asia	265	159
Africa	126	50
Australia	36	11
Australasia	20	8
Oceania	13	0
America	197	52
Antarctic islands	2	0

The *Saniculoideae* species are distributed in Asia, Europe, North Africa and the Middle East.

The species of the *Apioideae* subfamily are represented in the temperate zone of the Old World (Table 2).

2.3. CHARACTERS OF THE *CARUM* GENUS

2.3.1. General Description (Tutin *et al.* 1978)

Leaves 2–4 pinnate, sepals very small or absent. Petals whitish, rarely pink or yellowish, obovate, emarginate. Apex inflexed. Fruit obovoid-oblong, laterally compressed. Ridges filiform, prominent or almost winged. Vittae solitary. *Carum* genus is involved in the *Apioideae* subfamily, *Apieae* tribe and in *Seselinae* subtribe (Pimenov and Leonov 1993).

2.3.2. Species (Tutin *et al.* 1978, Hegi 1975)

It is written in the work of Engler (1964) that the *Carum* genus comprises about 25 species. Because of the great number of synonyms which usually mean different genera names as well, and the numerous subspecific taxa which are quite different in the works of different taxonomists, we describe only such taxa which can be delimited more or less clearly. Species with the greatest economic importance in the genus, *Carum carvi*, has been characterized in the next subchapter.

C. multiflorum (Sibth and Sm.) Boiss.
Biennial or perennial up to 70 cm. Basal leaves up to 10 cm, triangular in outline, 2–3 pinnate. Lobes up to 10 mm, ovate to obovate in outline. Rays 5–28, the outer is almost horizontal in fruit. Bracts and bracteoles 4–8, petals white. Fruit 2–3 mm, oblong-ellipsoid, ridges very narrowly winged. Distribution: South part of Balkan peninsula, and in South-East Italy.
    ssp. *multiflorum*: stems stout, with numerous branches. Rays usually 15–25. In West and South Greece, South Albania.
    ssp. *strictum* (Griseb)tin (syn.: *Bunium strictum* Griseb, *C. lumpeanum* Dörfler and Hayek). Stem slender, with few branches. Rays 5–12. Distribution: N.-E. Greece, S.W. Bulgaria, Central and North Albania, S.-E. Italy.

C. verticillatum (L.) Koch. (syn.: *Sison verticillatus* L., *Seseli verticillatum* Crantz, *Sium verticillatum* Lam., *Bunium verticillatum* Gren and Gordon, *Pimpinella verticillata* Jessen, *Apium verticillatum* Caruel, *Selinum verticillatum* E.H.L. Krause, *Aethusa fatua* Aiton, *Meum fatuum* Pers).
Erect, glabrous, perennial up to 120 cm. Root fleshy, stem striate, little branched, with few small leaves. Basal leaves 10–20 cm, usually oblong with more than 20 pairs of segments arranged in whorls. Rays up to 12, the number of linear-acuminate bracts is up to 10. The numerous bracteoles are usually deflexed. Petals white, fruit ellipsoid with

prominent ridges. Fruit size is 2.5–4 mm. Distribution: W. Europe, in marshes and damp meadows.

C. rigidulum (Viv.) Koch ex DC. (incl.: *C. graecum* Boiss and Heldr.,
C. adamovicii Halácsy).
Perennial, glabrous. Stem: erect, simple or with few and long branches and small leaves. Height of the stem is up to 60 cm. Basal leaves 10–20 cm, oblong, oblong-lanceolate, segments up to 15 pairs. Rays 3–11, erecto-patent. Bracts 0–6, linear. Bracteoles 3–8, linear, lanceolate, acuminate, scarious. Petals white or yellowish-white. Styles longer than stylopodium. The 3–4 mm long fruits have prominent ridges. Distribution: In mountains of Balkan peninsula and Central Italy.

C. heldreichii Boiss. (syn.: *C. flexuosum* (Ten) Nyman, incl.: *C. rupestre* Boiss and Heldr.).
Up to 40 cm, perennial. Stems several decumbent, later ascending, flexuous. Basal leaves 3–10 cm, oblong-lanceolate, segments up to 8 pairs, the largest near to the base of the lamina. Bracteoles 3–5, linear to setaceous, acuminate, with narrow scarious margin. Petals white or yellowish-white. Styles no longer than stylopodium. Fruit 3.5–4.5 mm, ellipsoid, ridges prominent. Distribution: In mountains of Greece, Albania and Italy.

2.3.3. *Carum carvi* L. (syn.: *Carum decussatum* Gilib., *Carum aromaticum* Salisb., *Carum officinale* S.F. Gray, *Apium carvi* Crantz, *Seseli carvi* Lam., *Seseli carum* Scop., *Ligusticum carvi* Roth, *Sium carvi* Bernh., *Bunium carvi* Bieb., *Foeniculum carvi* Link, *Pimpinella carvi* Jessen, *Selinum carvi* E.H.L. Krause, *Karos carvi* Nieuwland et Lunell, *Sium carum* Weber, *Aegopodium carum* Wibel, *Carvi careum* Bubani, *Pimpinella anisum* Meigen et Weniger nec L., *Lagoecia cuminoides* (Willemet ex DC.)nec. L).

2.3.3.1. Morphology (Tutin *et al.* 1978, Hegi 1975, Weberling 1981,
          Filarszky 1911, Corner 1963)

There are annual and biennial forms of *Carum carvi* (Hornok 1980). There are only very slight and uncertain differences in morphological and anatomical characters between these forms, so in the following we do not distinguish them.

Carum carvi has a taproot system. The main root is about 1 cm in diameter, weakly branched. It is whitish or brownish in color.

Stem divaricately branched, glabrous, perennial up to 150 cm. Stems striate, leafy (Figure 1). Leaves 2–3 pinnate, lobes 3–25 mm, linear lanceolate. The apex of the leaves are slightly dentate. Rays 5–16, unequal. Bracts usually absent, rarely up to 8 and then sometimes 2–3 partite. Bracteoles usually absent.

Flowers are united into an inflorescence forming a compound umbel. Flower is cyclic, with radial symmetry, usually heterochlamydeous, epigynous, pentamerous, tetracyclic. Calyx is reduced in size. Corolla comprises 5 whitish-green petals. Petals are three-lobed in outline, the apex of the central lobe is inflexed. Stamen are haplostemon and episepal, extrorz. Filaments are inflexed in the bud, and became erect only after

Figure 1   *Carum carvi*

Figure 2   Floral diagram of *Carum carvi*

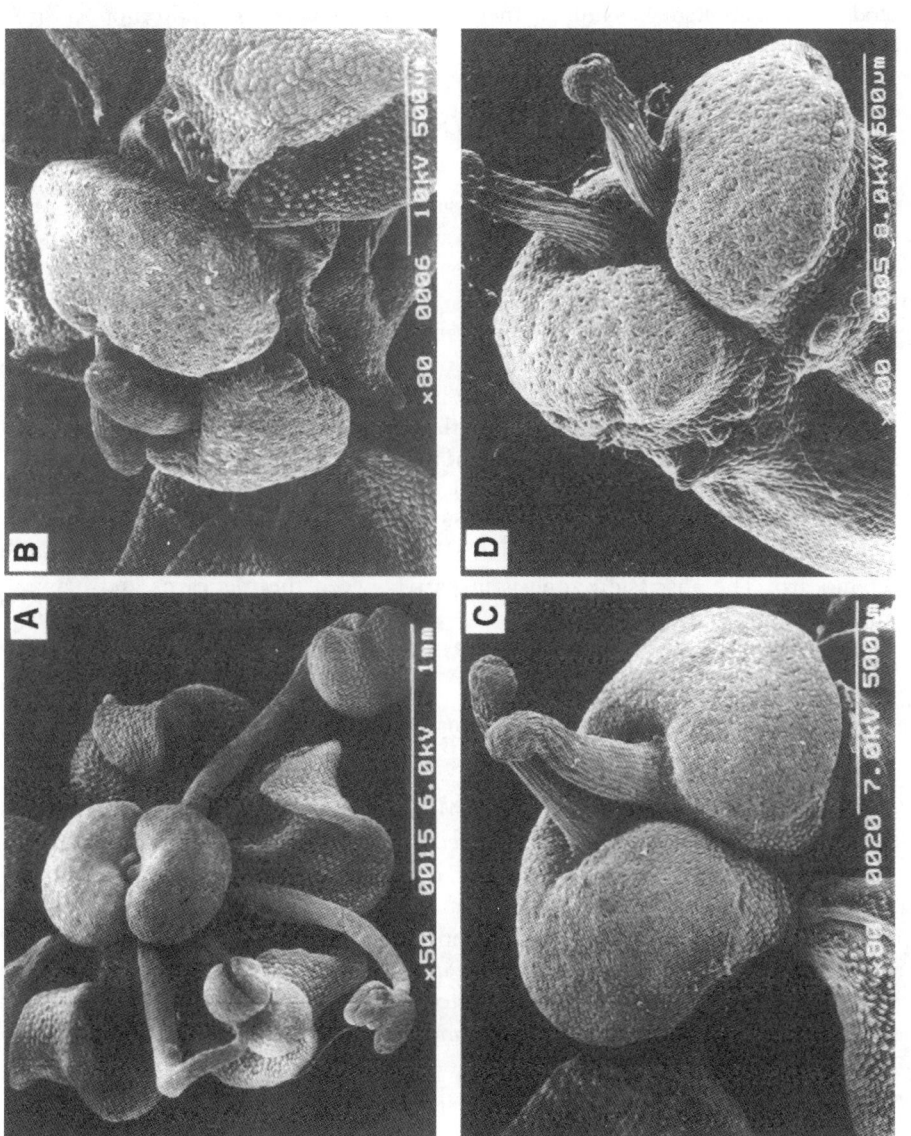

Figure 3 Development of the styles of *Carum carvi*. A: Young flower. B: Flower with developing styles. C: Flower with erect styles. D: Young fruit

splitting the anthers. The gynoecium consists of two carpels (Figure 2). On the top of the ovary and the young fruit there is a well developed discus divided into two more or less hemispherical parts. Styles originate from the upper-central part of the discus. Stigmata are round, with wet surface at the receptive phase.

Flowers are usually hermaphrodite but especially at the second half of the vegetation period and in the lateral smaller umbels there are male flowers too (andromonoecism).

By *Carum* we can find proterandry-dichogamy. It means, that the anthers ripe and dehisce before the stigma becomes receptive (Figure 3A). The size and the spatial arrangement of the styles change during maturation. At the time of the splitting of the anthers styles are visible as small projections of the two large discus (Figure 3A). In the following step styles are larger, laying on the surface of the discus parallel with each other (Figure 3B). Before the receptive phase of the stigmata, styles become erect (Figure 3C) and at the receptive phase the two styles are standing in V-form. It is visible on the young fruit too (Figure 3D). The styles dry out, turn down and remain on the top of the mericarps after maturation.

The fruit is dry and indehiscent, the type of it is cremocarp. Fruit 3–6 mm, ovoid, with special smell. Ridges low, rounded. Each mericarp has five ribs with vascular bundles in them. The mericarps develop from a single carpel of the inferior ovary and separate form one another along the adnation but they remain connected with to the forked carpophore. The origin of the carpophore is not clear. According to the opinion of different anatomists it may arise from flower axis or from the carpels. It seems, that the different areas of the caropohore differ in origin. The basal portion is of receptacular origin while the upper area is carpellary and contains the two ventral vascular bundles of the cremocarp. The abscission zone is in part between the mericarps and between the mericarp and the carpophore.

2.3.3.2. *Anatomy* (Fahn 1982, Esau 1969, 1977, Johri et al. 1992, Eames and McDaniel 1951, Korsmo 1954)

2.3.3.2.1. *Vegetative organs*

Root: The secondary thickened root (Figure 4) is covered by periderm. Cork derived from the phellogen has 4–6 regularly arranged cell layers of quadrangular cells elongated periclinally. Under the phellogen there is a pericyclic parenchyma, the phelloderma ring with numerous parenchyma cells and wide intercellular spaces among them. The pericyclic parenchyma merges almost imperceptibly with the phloem. The small sieve elements and the companion cells are arranged in groups and can be distinguished from parenchyma by their size and darker cytoplasm. In the periphery of the phloem bundle the sieve elements collapse and are almost invisible. There are numerous small oil cavities in the phloem. The epithel cells of these cavities are thin walled and are arranged in one row. The transversally dilatated cells of the phloem rays are filled with large starch grains.

The cambium is practically invisible. The secondary xylem is not delimited with a well defined ring of cambium. The vessels are different in diameter. The vessels are accompanied with parenchyma cells and some xylem fibers. The xylem is divided into sectors by the wide parenchyma rays. The central area of the secondary xylem comprises

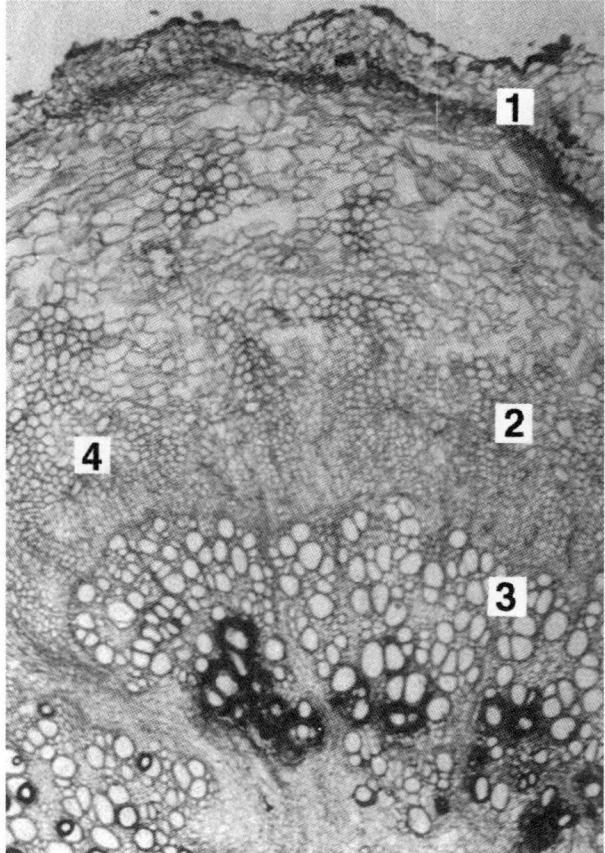

Figure 4 Secondary root cross section. 1: periderm, 2: phloem, 3: xylem, 4: oil ducts

parenchyma with thin cell walls. In the peripherial ring, near to the cambium, parenchyma of thick lignified walls predominates. In the older (central) area of the secondary xylem there is a dark coloration in some xylem parenchyma cells around vessels. Dark material accumulated inside these vessels too. Such coloration never occurs in the younger secondary xylem.

The primary xylem is excentric in its position and comprises many parenchyma cells. In the primary state the root shows the tissue structure of the typical tetrarch root.

Stem: The young stem is ribbed and has a pith of parenchymatic cells. These cells are broken in the older stem and a pith cavity occurs. Parallel with it, ribs become smaller or almost disappear. The stem is covered with epidermis (Figure 5). The outer tangential wall of the epidermis is thicker and covered by cuticle. Under the epidermis there are collenchyma bundles in the ribs. In older portion of the stem they are visible also in the valleculae. Between the collenchyma bundles, just beneath the epidermis there are 3–4 cell layers of chlorenchyma with cells rich in chloroplasts. Stomata are only in the epidermis above chlorenchyma. Under the stomata there are air chambers in

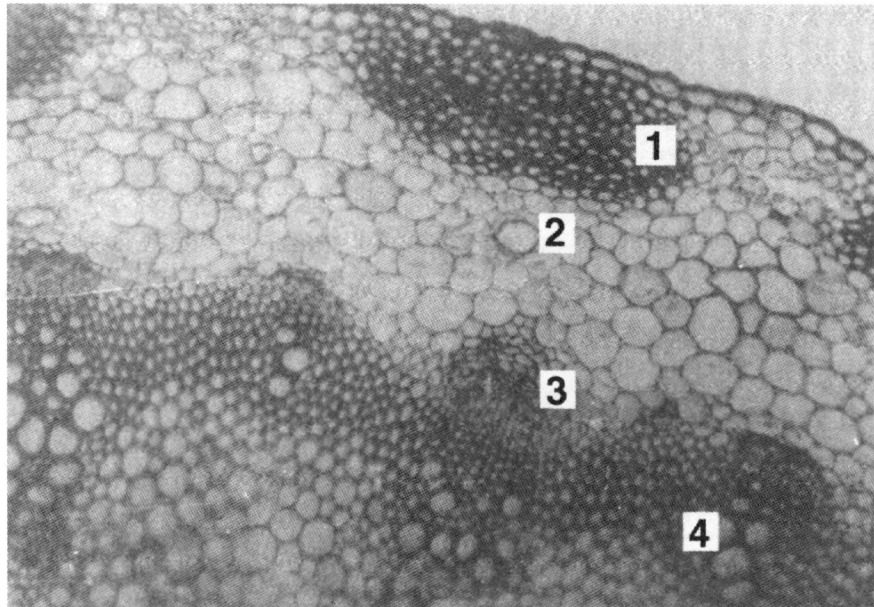

Figure 5   Stem cross section. 1: sclerenchyma bundle, 2: oil duct, 3: phloem, 4: xylem

the chlorenchyma. The cells of primary cortex are larger than that of the chlorenchyma. Their diameter is growing towards the vascular tissues. Cortical parenchyma cells are thin walled, almost colorless in their protoplasma, loosely arranged, with large triangular or quadrangular intercellular spaces among them. In the cortex there are schizogenous secretory ducts surrounded by epithel tissue. In the cells of epithel, which are much more smaller than that of the primary cortex, there are large nuclei and dense protoplasma. There are also some idioblasts in the primary cortex filled with crystal sand. In the stem of *Carum* there is no anatomical delimitation of the primary cortex (starch sheet). The collateral vascular bundles comprise one wavy ring. The vascular bundles are oval in outline. Cambial activity is not high. The secondary vascular tissue can be well distinguished in the xylem: in the secondary part of the xylem the vessels are much more smaller, surrounded and separated by fibers (parenchyma cells with thick lignified walls). The primary part of the bundle comprises vessels of relatively wide diameter. The xylem bundles are separated from each other in this area by parenchyma of thin walls.

The phloem is more or less hemispherical in shape, with weakly developed phloem fiber cap on its peripherial side. In most cases the oil cavities are near to the phloem.

Leaf: The leaf blade is thin, covered by unicellular epidermis. There is considerable difference in the size of the epidermis cells in the adaxial and abaxial surface. The adaxial cells are larger than the abaxial ones. Epidermis is covered by thick cuticle. The ornamentation of the cuticle is wawy-striped (Figure 6). Leaves are bifacial (Figure 7), with one layer of palisade parenchyma and 5–6 layers of spongy parenchyma. Stomata (mesomorf) occur on the both epidermis (amphistomatic type) and at the margin of the

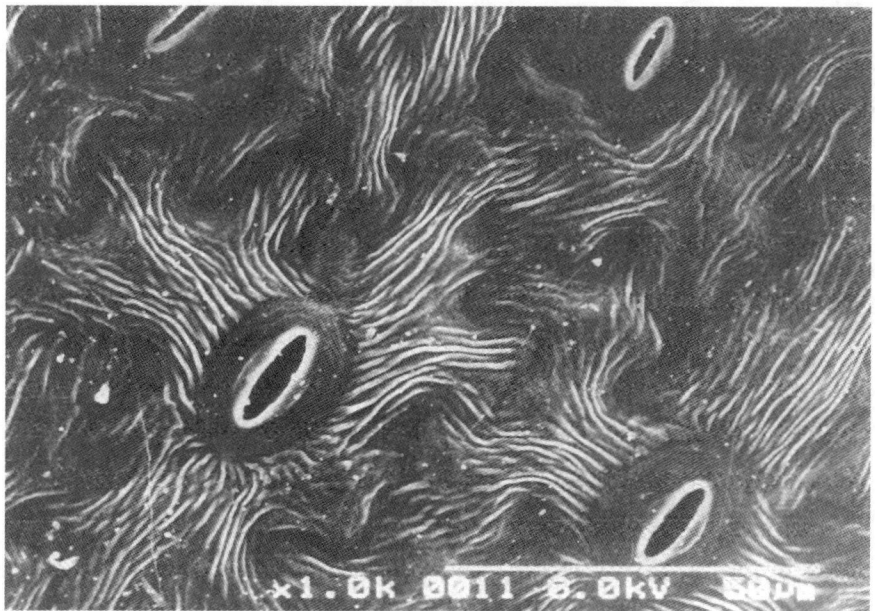

Figure 6 Lower epidermis of the leaf of *C. carvi*

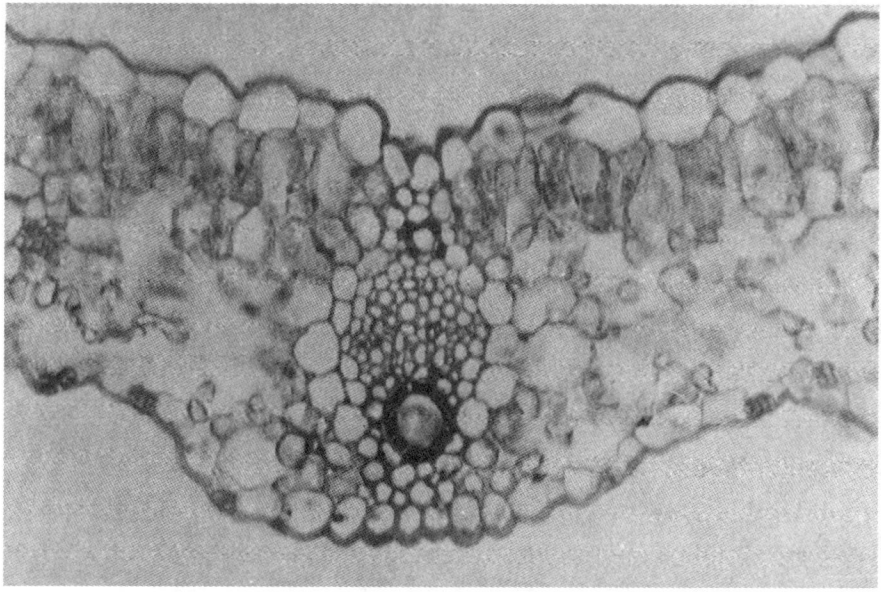

Figure 7 Leaf cross section with oil ducts associated to the midrib

leaves. Stomata are anomocytic. The central vascular bundle of the leaves is in a rib. To the adaxial and abaxial side of the collateral vascular bundle there are attached two oil ducts (Figure 7). The adaxial oil duct is smaller, the abaxial one is about twice as large. The epithel cells of the oil cavities show dark coloration. In the vascular bundles the xylem vessels are arranged in one or two rows, in the larger bundles more or less parallel with the leaf surface.

The lateral vascular bundles of the leaves are smaller and are connected only with one oil duct at the abaxial side. Vascular bundles are surrounded by parenchymatic bundle sheath.

The petiole comprises 8 oval, collateral vascular bundles. It is U-shaped in cross section, covered by the epidermis. The rigidity of the petiole is increased by the collenchyma bundles just beneath the epidermis.

2.3.3.2.2. Generative organs

Flower: The calyx of the Carum flower is very small. It is covered by thin walled epidermis. The mesophyll is homogenous, with a single vascular bundle in central position.

The corolla comprises 5 free petals. Petals are attached to the floral axis with narrow base. The upper epidermis of the petal has papillae with parallel cuticular ornamentation (Figure 8). The mesophyll is homogenous with some vascular bundles.

The stamen have long filaments. The filaments attach the dorsal side of the anthers. The thecas have extrorz opening in the form of a longitudinal split. The endothecium has parallel wall thickenings. The development of the anther wall is dicotyledonous type on the basis of the formation of the middle layer. It means, that in the young anther the hypodermal

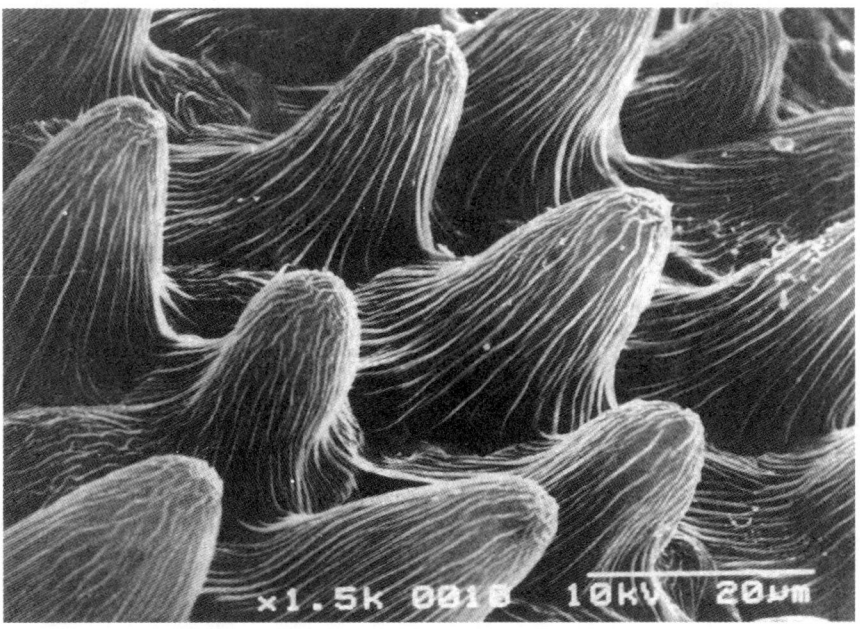

Figure 8 Papillae on the petal

archesporial cells divides periclinally. The outer cells form the primary parietal layer, the inner one is the primary sporogenous tissue. The primary parietal layer bears two secondary parietal layers. The peripheral one divides into the endothecium, and the middle layer, while the central one is functioning as the tapetum. The type of the tapetum is secretory or glandular. The cytokinesis of the pollen mother cells is simultaneous type. The thick callose wall intrudes centripetally and separates the four microspores only after the second division of the pollen mother cells. Pollen tetrads are tetrahedral, decussate or isobilateral.

The tricolporate pollen grains have bilateral symmerty (Figure 9). At the time of shedding they are at three-celled state.

The gynoecium fused of two carpels. The upper part of the ovary is surrounded by a nectary (discus). On the surface there are numerous nectar stomata sunken below the epidermis with special ornamentation on its cuticle (Figure 10). The styles are covered with axially elongated epidermal cells. The surface of the stigma is spherical, stigmatic papillae does not occur. The type of sitgmata surface is wet.

The ovary has two cavities with 1–1 ovules in them. The ovary wall is parenchymatous. At the apical part of the ovary, at the level of the discus the ovary wall consists of small isodiametric cells (Figure 11). The epidermis and the subepidermal layers are dense in their protoplasts. These cells are involved in the nectar production of the discus. There are oil canals in this area of the ovary too. Canals are associated with vascular bundles, ribs and valleculae can not be distinguished.

The ovule is anatropous. Hypostase differentiates parallel with the archesporial state, obturator develops too. The ovule is suspended from the upper area of the ovary at the

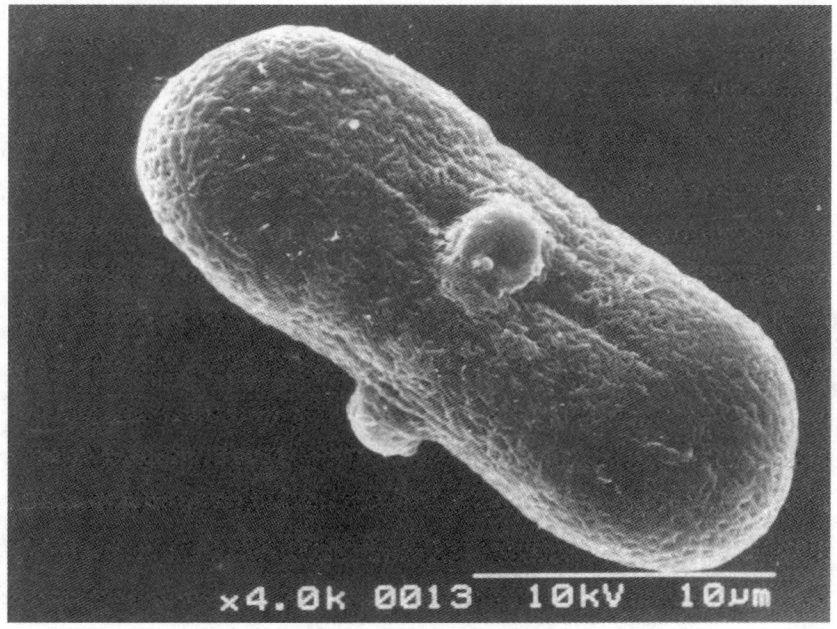

Figure 9 Pollen grain of *C. carvi*

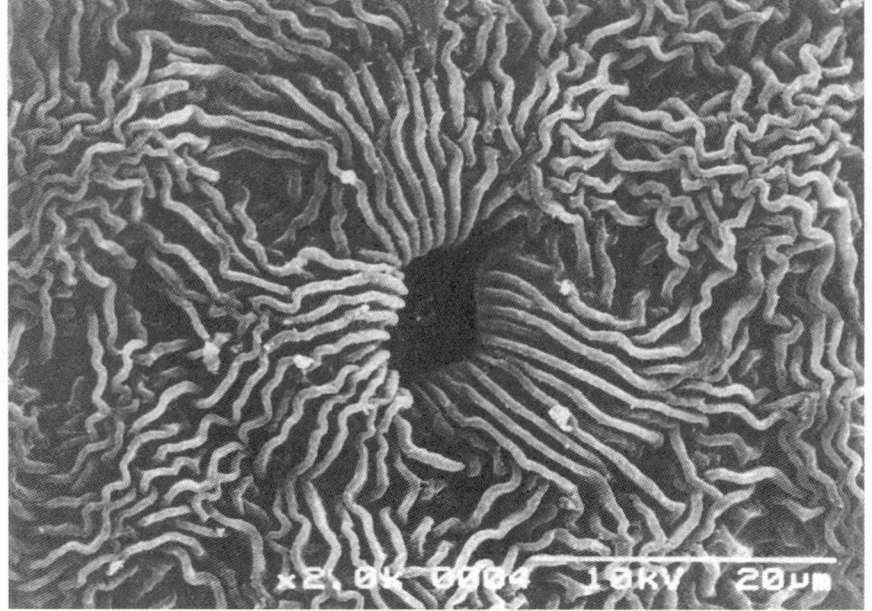

Figure 10　Surface of the discus with nectar stoma

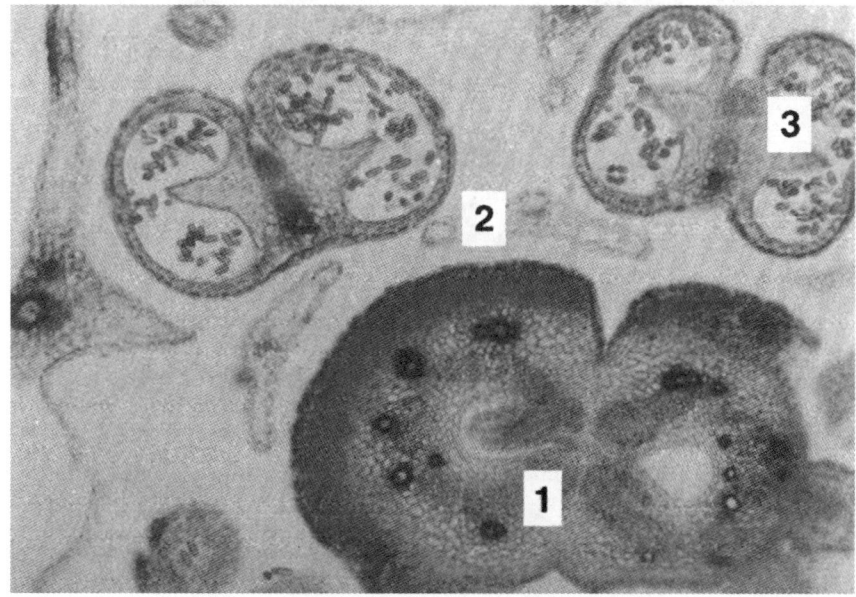

Figure 11　Cross section of the apical part of the ovary. 1: ovary, 2: sepal, 3: anther

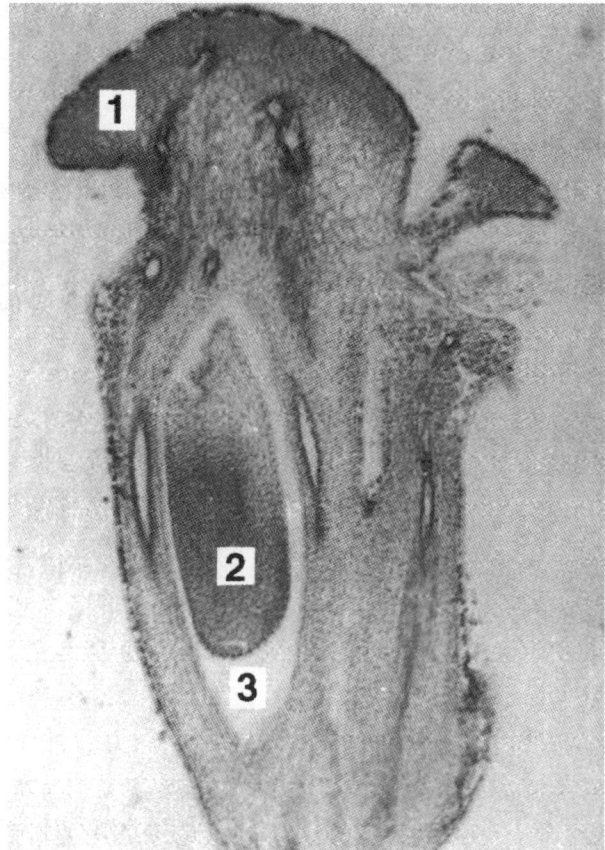

Figure 12   Longitudinal section of the young ovary. 1: discus, 2: ovule, 3: cavity of the ovary

commissural side. The micropyle is situated on the outer and upper part of the anatropous ovule. In young state the ovule does not fill the cavity of the ovary (Figure 12). The single integument is well visible. Figure 12 illustrates, that the abortion of one carpel does not influence the development of the other. After fertilization the ovule is growing intensively in its length and fills the whole cavity of the ovary.

In the cross section of the ovary (Figure 13) it can be detected the body of the nucellus with the embryo sac in it. The embryo sac develops from the hypodermal archesporial cell or from the chalazal cell of the tetrad. The tetrad may be T-shaped. The embryo sac is surrounded by the endothelium (integumentary tapetum), the innermost cell layer of the integument. In the fruit wall there are oil ducts in the valleculae and in the ridges associated with the vascular bundles. In this developmental step these two types of oil ducts are almost the same in diameter.

Around the embryo sac the cell walls of the nucellus disappear, the embryo sac "digests" and absorbs the tissue of the nucellus (Figure 14). The absorbed material is

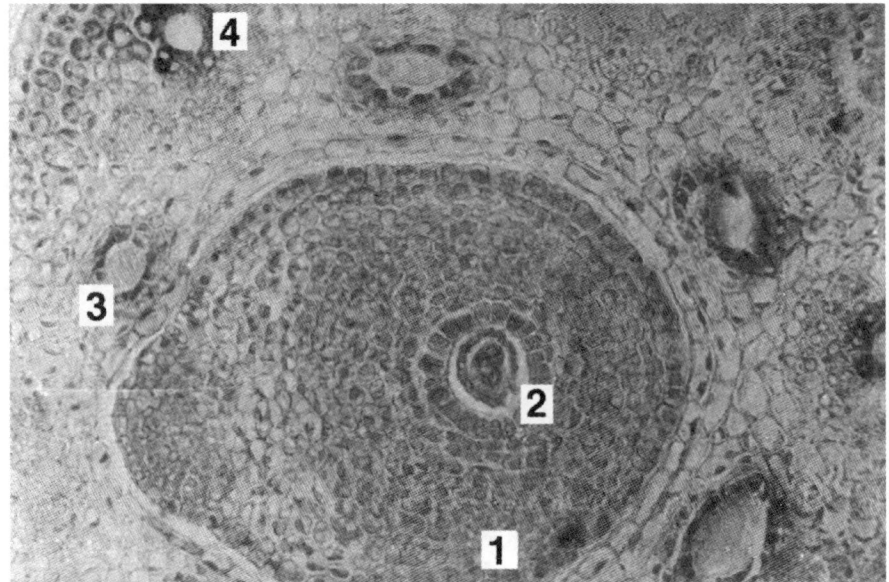

Figure 13 Cross section of the ovary. 1: nucellus, 2: embryo sac, 3: vallecular oil duct, 4: carinal oil duct

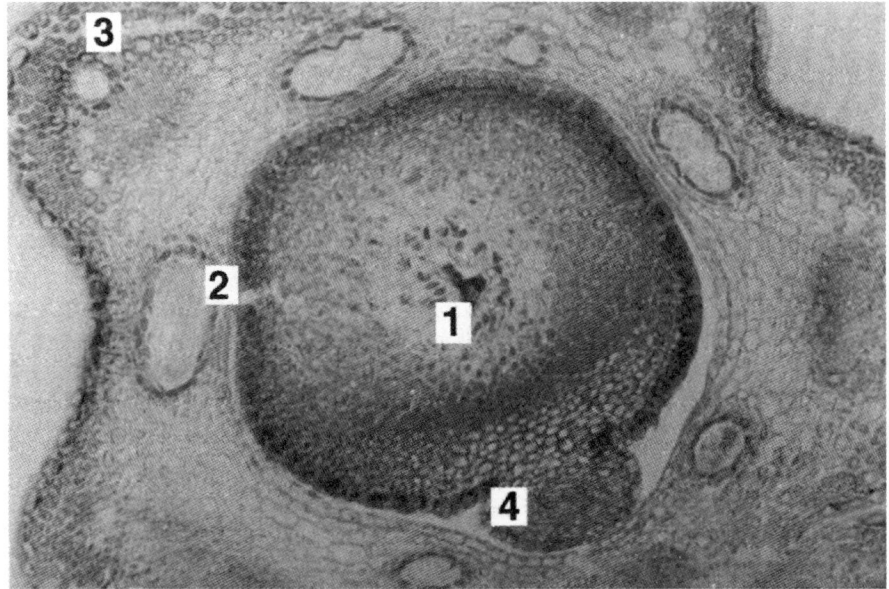

Figure 14 Developing ovule. 1: embryo sac, 2: vallecular oil duct, 3: carinal oil duct, 4: funiculus

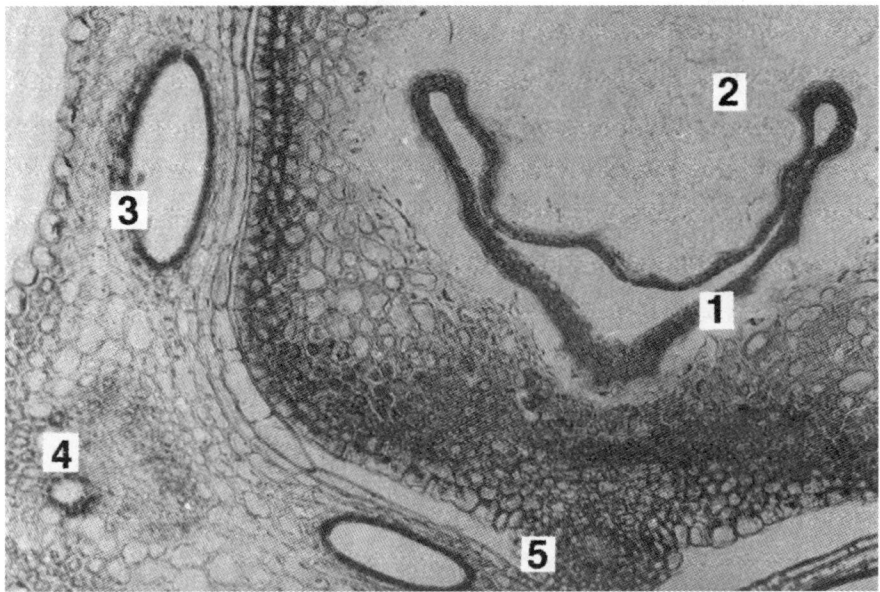

Figure 15 Developing embryo sac, the cellular endosperm emerge. 1: cellular endosperm, 2: digested nucellar tissue, 3: vallecular oil duct, 4: carinal oil duct, 5: funiculus

incorporated into the endosperm. The endosperm is nuclear. The first nuclei consist of a group at the micropylar region. Wall formation starts after 64 or more nucleate state. Parallel with this phenomenon, in the fruit wall the areas of the vallecular oil ducts increase. They are much more larger, than ducts in the ribs.

In the following period the embryo sac digests the great portion of the nucellus, and along the wall of the embryo sac emerges the first layer of cellular endosperm. The areas of the vallecular oil ducts are growing (Figure 15). In longitudinal section (Figure 16) we can follow the digestion of the nucellus from the apex of the young seed towards the base. The voluminous nuclear endosperm (nuclei are invisible) is surrounded by the thin layer of cellular endosperm what is broken in the left seed. In this developmental step the upper half of the carpophor has already been splitted.

The diameter of the vallecular oil ducts (and probably the amount of the volatile oils in them) are the largest when the whole endosperm becomes cellular. The central area of the endosperm is digested by the growing embryo (Figure 17). The embryogeny is of the Solanad type. In some cases, besides the small and straight zygotic embryo, development of synergid proembryo can be observed, which is much more smaller than the zygotic one (Figure 18).

Seed and fruit: The seed is endospermic, the embryo is small. A single layer of the epidermis represents the seed coat. The raphe of the seed is ventral. The reserve material of the thick walled endosperm is oil and numerous aleuron grains (Figure 19).

The form of the endosperm in the seed has diagnostic value. By Carum carvi the endosperm is bulging at the area of the raphe in cross section (Figure 20). It means that

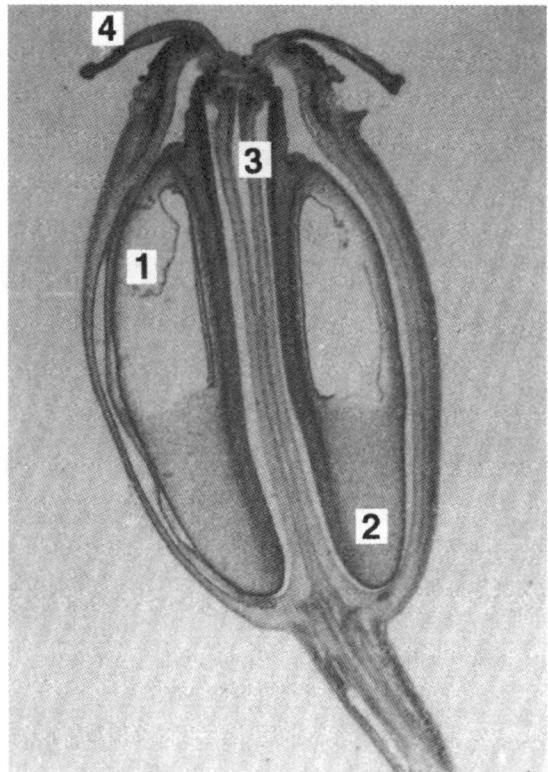

Figure 16 Longitudinal section of the young fruit. 1: embryo sac, 2: nucellus, 3: carpophore, 4: style

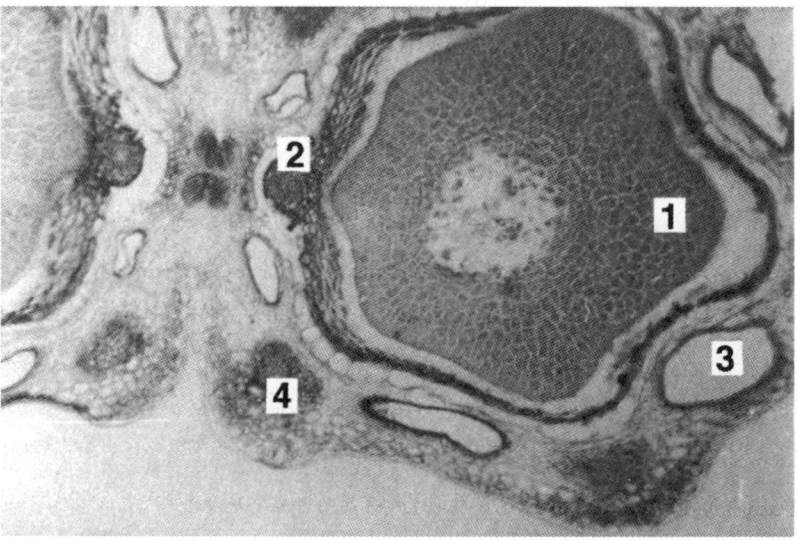

Figure 17 Cross section of the young fruit. 1: cellular endosperm, 2: funiculus, 3: vallecular oil duct, 4: carinal oil duct

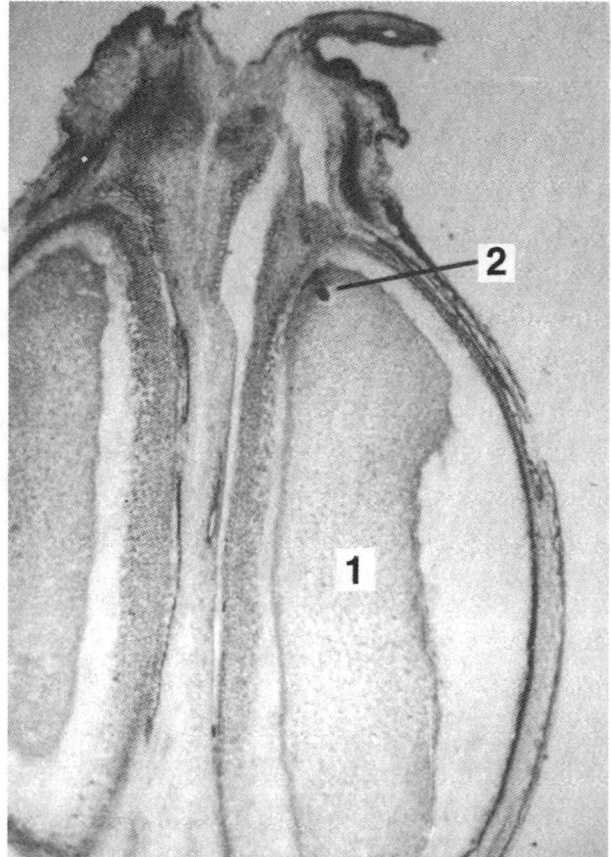

Figure 18 Proembryo in the seed. 1: cellular endosperm, 2: proembryo

this species is of the orthospermae type. In other species the surface of the raphe is U-shaped (campylospermae type) or convex (coelospermae type). These characters are used in classification of the Umbelliferae species but they have no phylogenetic importance.

The fruit is schizocarp (cremocarp). The exocarp has small isodiametric cells. The ornamentation of the cuticle is parallelly-striped (Figure 21). There are stomata on the surface of the fruits. The mesocarp is parenchymatous, 8–11 cell layers in thickness. The endocarp (inner epidermis of the fruit wall) is unicellular. It consists of narrow cells with different spatial arrangement (parallel or perpendicular to the transverse axis of the mericarp). The mesocarp of the fruit becomes thin during maturation because of dual purpose: it is drying out, and has been pressed by the great mass of endospermium. The areas of vallecular oil ducts also decrease.

The carpophore splits at maturation as a result of an abscission zone. This zone is a lignified tissue between the two ventral veins (Figure 22).

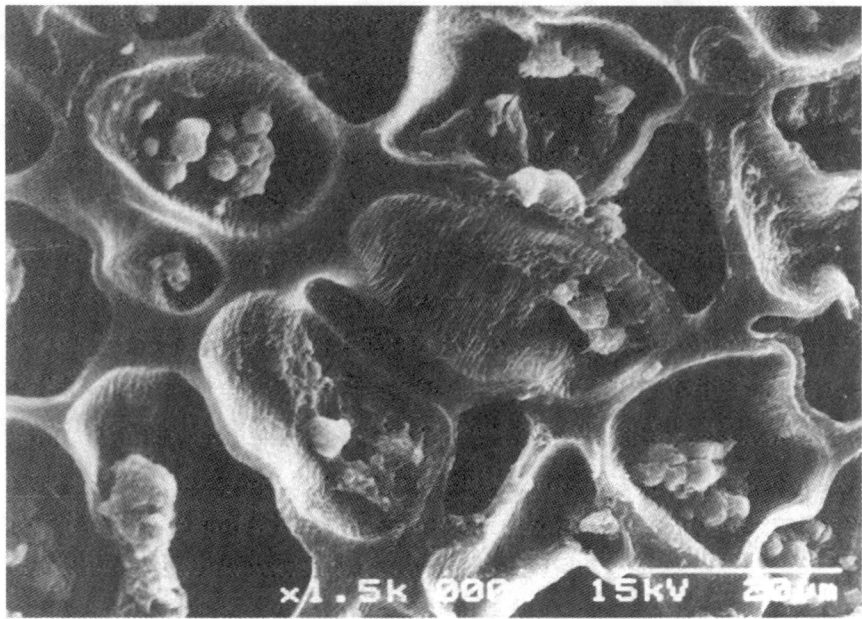

Figure 19 Endosperm cells with aleuron grains

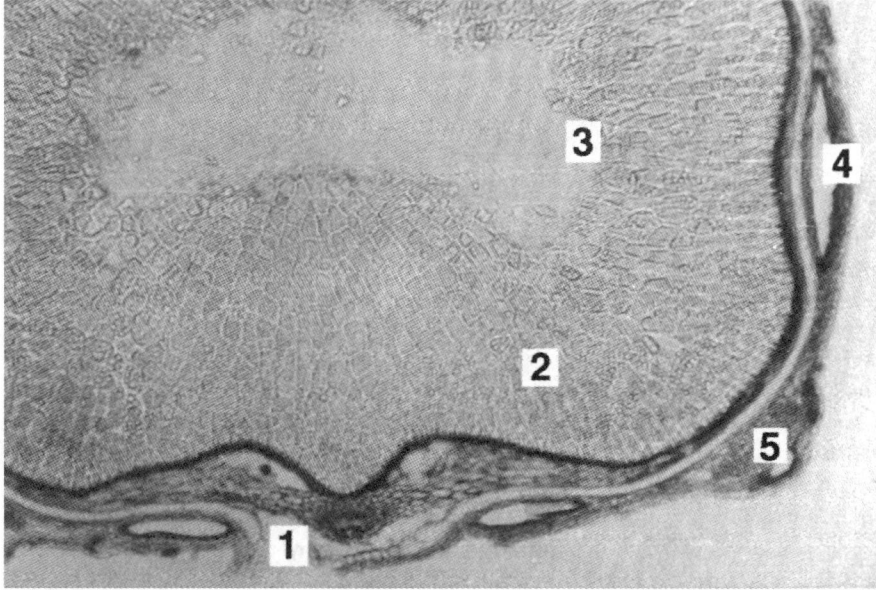

Figure 20 Cross section of the mature seed. 1: raphe, 2: endosperm, 3: cavity of the embryo with digested endosperm cells, 4: vallecular oil cavity, 5: carinal oil cavity

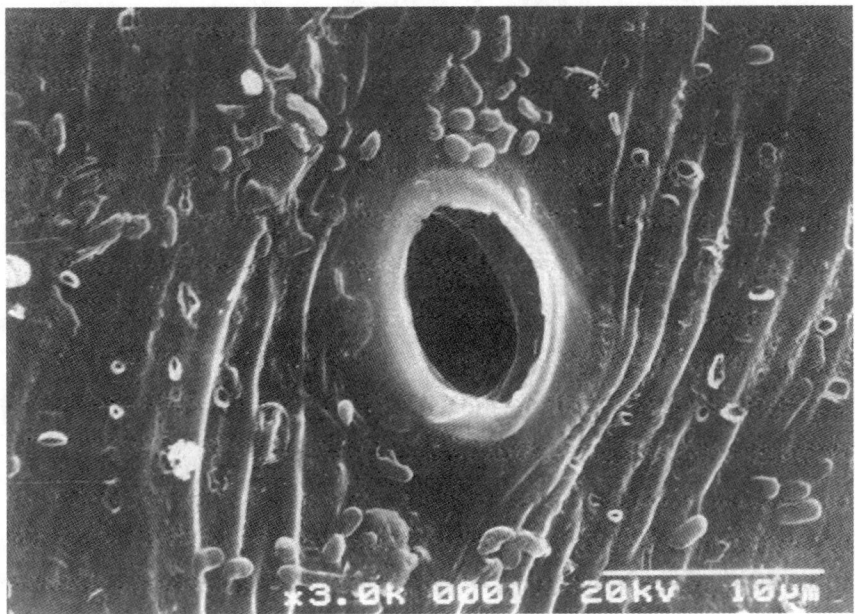

Figure 21   Stoma of the fruit wall

Figure 22   Schizocarp

2.3.3.3. Subspecific Division and Geographical Distribution of C. carvi

Numerous forms of *Carum carvi* can be distinguished on the basis of morphological differences, especially on the shape, number, persence or absence of bracts and bracteoles. In the monograph of Hegi (1975) there are the following 10 forms: f. *nanum* DC., f. *demissum* Murr, f. *coarctatum* (Tinant)Baguet, f. *latisectum* Thellung, f. *vulgare* Alef, f. *intermedium* Rouy, f. *pterochlaenum* DC., f. *proliferum* Peterm, f. *alpinum* Schur, f. *atrorubens* Schube.

Soó (1966) describes 9 of the above form, with the exception of f. *coarctatum*.

Carum carvi is distributed especially on different areas of the North hemisphere. It is the species of the natural flora in North and Central Europe, England, East and Central French, South Spain, North Italy, Balkan peninsula, Central Asia. It is spread also as a result of human activity, especially in Holland, North Africa, North America and New Zeland.

ACKNOWLEDGEMENT

I express my special thanks to Erika Nagy, Anna Medvegy and Katalin Kálmán for their technical assistance.

REFERENCES

Bell, E.A. and Charlwood, B.W. (1980) *Secondary Plant Products.* Springer Verlag Berlin, Heidelberg, New York.

Bohlmann, F. (1971) Acetylene compounds in the *Umbelliferae*. In Hegnauer, R. (ed.) *The biology and chemistry of the Umbelliferae.* Acad. Press, London.

Borhidi, A. (1995) *A Zárvatermők Fejlődéstörténeti Rendszertana.* Nemzeti Tankönyvkiadó, Budapest.

Corner, E.J.H. (1963) *The Seeds of Dicotyledons.* Cambridge Univ. Press, Cambridge.

Cronquist, A. (1968) *The Evolution and Classification of Flowering Plants.* The New York Bot. Gardens Bronx, New York.

Crowden, R.K., Harborne, J.B. and Heywood V.H. (1969) Chemosystematics of the *Umbelliferae* – a general survey. *Phytochemistry* 8: 1963–1984.

Dahlgren, K.V.O. (1980) A revised system of classification of the *Angiosperms. Bot J. Linnean Soc.* 82: 89–92.

Drude, O. (1897) *Umbelliferae.* In Engler, A. and K. Prantl, (eds.) *Die Natürlichen Pflanzenfamilien* 83: 63–250.

Eames, A. and McDaniels, L. (1951) *An Introduction to Plant Anatomy.* McGraw Hill Publ. Co., New York, London, Toronto.

Engler, A. (1964) *Syllabus der Pflanzenfamilien.* Gebrüder Borntraeger.

Esau, K. (1969) *Pflanzenanatomie.* Fischer Verlag, Stutgart.

Esau, K. (1977) *Anatomy of seed plants.* John Wiley and Sons, New York.

Fahn, A. (1982) *Plant anatomy.* Pergamon Press Oxford, New York.

Filarszky, N. (1911) *Növénymorphologia.* Franklin Társulat, Budapest.

Guyot, M. (1971) Phylogenetic and systematic value of stomata of the *Umbelliferae*. In Hegnauer, R. (ed.) *The biology and chemistry of the Umbelliferae.* Acad. Press, London.

Hegi J. (1975) *Illustrierte Flora Mitteleuropa* Band V. Teil 2. Paul Parey Verlag, Hamburg.

Hegnauer, R. (1971) Chemical patterns and relationships of Umbelliferae. In Hegnauer, R. (ed.) *The biology and chemistry of the Umbelliferae.* Acad. Press, London.

Hegnauer, R. (1964) *Chemotaxonomie der Pflanzen.* Springer Verlag, Berlin.

Heywood, V.H. (1971) Systematic survey of Old World *Umbelliferae.* In Hegnauer, R. (ed.) *The biology and chemistry of the Umbelliferae.* Acad. Press, London.

Hornok, L. (ed.) (1990) *Gyógynövények Termesztése és Feldolgozása.* Mezőgazdasági Kiadó, Budapest.

Hutchinson, J. (1969) *Evolution and Phylogeny of Flowering Plants.* Acad. Press, New York.

Johri, B.M., Ambegaokar, K.B. and Srivastava, P.S. (1992) *Comparative Embryology of Angiosperms.* Vol. 1. Springer Verlag, Berlin.

Kaul, B. and Staba, E.J. (1967) *Ammi visanga* tissue cultures: Multilitre suspension growth and examination furanochromones. *Planta Med.* **15**: 145–156.

Klischies, M., Stockigt, J. and Zenk, M.H. (1975) Biosynthesis of the allylphenols eugenol and methyl-eugenol in *Ocimum basilicum.* Chem. Commun. 879–880.

Korsmo, E. (1954) *Anatomy of weeds.* Kristes Booktrykkeri, Oslo.

Melchior, H. (1964) *Umbelliflorae.* In Engler, A. *Syllabus der Pflanzenfamilien.* 12 Aufl. Bd. 2. Gebrüder Borntraeger, Berlin.

Pickering, J. and Fairbrothers, D.E. (1971) The use of serological data in a comparison of tribes in the *Apioideae.* In Hegnauer, R. (ed.) *The biology and chemistry of the Umbelliferae.* Acad. Press, London.

Pimenov, M.G. and Leonov, M.V. (1993) *The genera of the Umbelliferae.* Royal Botanic Gardens, Kew.

Schmitz, M. and Seitz, U. (1972) Hemmung der Anthocyaninsynthase durch Gibberellinsaure A3 bei Kalluskulturen von Daucus carota. *Z. Pflanzenzüchtung* **68**: 259–265.

Soó, R. (1966) *A magyar flóra és vegetáció rendszertani és növényföldrajzi kézikönyve.* Vol. II. Tankönyvkiadó, Budapest.

Takhtajan, A. (1969) *Flowering plants. Origin and dispersal.* Oliver and Boyd, Edinburgh.

Tutin, T.G., Heywood, V.H., Burges, N.A., Moore, D.M., Valentine, D.H., Walters, S.M. and Webb, D.A. (1978) *Flora Europea* Vol. 2. Rosaceae to Umbelliferae. Cambridge University Press, Cambridge.

Weberling, F. (1981) *Morphology of flowers and inflorescences.* Cambridge University Press, Cambridge.

3. MAIN CHEMICAL CONSTITUENTS OF *CARUM*

JOANNA RUSZKOWSKA

The Faculty of Chemistry, University of Warsaw,
ul. Pasteura 1, 02–093 Warszawa, Poland

3.1. INTRODUCTION

Oleum Carvi, an essential oil obtained by steam distillation of ground caraway [*Carum carvi* (L)] seed, has been mentioned in the textbook of pharmacology more than 120 years ago (Nathnagel and Rossbach 1883). Already then the main constituent of this oil was characterized as a compound with molecular formula $C_{10}H_{14}O$, isomeric with tymol, having similar chemical properties to some constituents of terpentine oil. Although not all of the above statements occurred to be true, the monoterpene was named carvol after the name of the plant it was first isolated from in 1841. Constant development of analytical and isolation methods enabled organic chemists to characterize precisely most of the individual constituents of *Carum carvi*, the more so still appearing new aspects of its possible application and utilisation stimulated new interests.

Once having the analytical methods of determining and tracing of individuals in the plant established, it became possible to inspect the influence of various factors upon its chemical composition. So, the dependence of chemical composition upon such factors as agricultural treatment, isolation method and storage conditions was in many cases established in details.

The other modern trend in plant analysis is bioassay-guided fractionation of natural extracts. It is interesting to note that in the case of *Carum carvi* essential oil the search for compound(s) inducing the detoxyfication process of enzyme glutathione S-transferase (GST) in several mouse target tissues has led again to S(+)-carvone (formerly carvol) and limonene, identified by means of ^{1}H NMR, IR and EI MS spectroscopy. According to Zeng Guo Quiang *et al.* (1992), carvone exhibited the highest activity as GST inducer among compounds under study thus revealing its interesting possible anti-carcinogenic activity.

Chemical constituents of a plant are classified as primary and secondary metabolites. And however the first group comprises substances playing a vital role and necessary in normal cell life processes, the second is usually of broader interest. This is especially the case of pharmacologically useful plants with secondary metabolites as bioactive substances.

From pharmacological point of view the most interesting substances of *Carum carvi* are terpenes – constituents of caraway essential oil. It is usually obtained by a steam distillation of seeds with subsequent extraction and solvent removing. *Oleum carvi* has found many applications as anticovulsant, antiflatulent, lactogenic, carminative agent and vehiculum. The growing attention is paid nowadays to phenolic constituents of *Carum* due to their antioxidative properties.

The main groups of chemical constituents of *Carum carvi* discussed in necessary details in next paragraphs of this chapter, are the following:

– primary metabolites: saccharides, non-terpenoic lipids, aminoacids and peptides, organic acids, and
– secondary metabolites: terpenes, flavonoids, coumarins and phenolic compounds.

Some attention will be devoted also to trace elements and "unwanted guests" – components, detected in caraway grown or cultivated in polluted areas or resulted from an inappropriate chemical treatment of plants or seeds during vegetation period, processing or storage.

3.2. GENERAL CHARACTERISTICS OF CARAWAY RAW MATERIAL

Moisture, ash, lipids, proteins, fiber, essential oil, total sugars and starch contents in fresh or dried plant samples are usually parameters that should be determined in order to characterize raw material of a given pharmacognostic value. These data are usually determined according to methods described in A.O.A.C. (A.O.A.C. 1975). Determination methods of physico-chemical properties of *Carum* essential oil (specific gravity, optical rotation, refractive index, boiling range, evaporation residue as well as acid value, ester number, alcohol content, iodine number or peroxide value) are described in Pharmacopeias (for example Polish Pharmacopeia IV, British Pharmacopeia 1973, etc).

The content of given fraction of constituents in raw material may differ to great extent, as can be seen on the basis of data presented according to El Wakeil *et al.* (1986a), Abou-Zied *et al.* (1974), Stepanenko *et al.* (1980) and others in the Table 1.

Table 1   Chemical Compounds of *Carum* – General Characteristics

Group of constituents	Lowest content [% of d.wt.] (Ref.)	Highest content [% of d.wt.] (Ref.)	Average content [% of d.wt.] (Ref.)
Essential oil	0.99 (El Wakeil 1986a)	8.1 (Atal 1967)	5.5 (Stepanenko 1980)
Saccharides (total)	7.3 (Abou-Zied 1974)	8.5 (El Wakeil 1986a)	
N-containing compounds (total) starch	20.0	29.9 (Abou-Zied 1974)	25.0–27.0 (El Wakeil 1986a) 24.0 (El Wakeil 1986a)
Lipids	2.6 (Abou-Zied 1974)	12.6 (El Wakeil 1986a)	8.0–10.0 (Stepanenko 1980)
Ash	7.7 (El Wakeil 1986a)	10.0 (Khan 1985)	
Moisture	5.0 (El Wakeil 1986a)	7.5 (Schulz 1980)	
Fibers			22.3 (El Wakeil 1986a)

Discrepancies between the contents of given fraction of active compounds determined by cited authors may reflect *Carum* varietal differences, site and vegetation period effects and may derive from specific methodics applied.

3.3. PRIMARY METABOLITES OF *CARUM CARVI*

The following main primary metabolites are identified and characterized as a result of caraway samples extraction/separation procedures: sugars (mono-, oligo- and polysaccharides), lipids, aminoacids, proteins, proteids, free organic acids. They are spread uniformingly within all tissues of the plant.

3.3.1. Saccharides of *Carum carvi*

Mono-, oligo- and polysaccharides found in all parts of the plant serve as a temporary reserve material. The monosaccharides identified in caraway fruits, leaves and spare tissues are hexoses: glucose and fructose. Main reserve disaccharide is saccharose (sucrose), found in quantity 1% and 3% of fresh plant weight in fruits and leaves respectively. The changes in main mono- and disaccharides content during caraway two years vegetation period is described in details by Hopf and Kandler (1976). Other disaccharides found in the seeds in minor amounts are: trehalose, glucosyl mannose and mannityl-1-β-glucose.

The most interesting trisaccharide of *Carum* is perhaps umbelliferose (Figure 1a), an isoraffinose typical to umbellifers, found in all parts of the plant and serving as a temporary reserve material similar to sucrose. It occurs in greater amounts than sucrose only in the ripe fruits and is not preferentially accumulated in any particular vegetative organ. Physiology of this sugar, its translocation and turnover was studied in details by

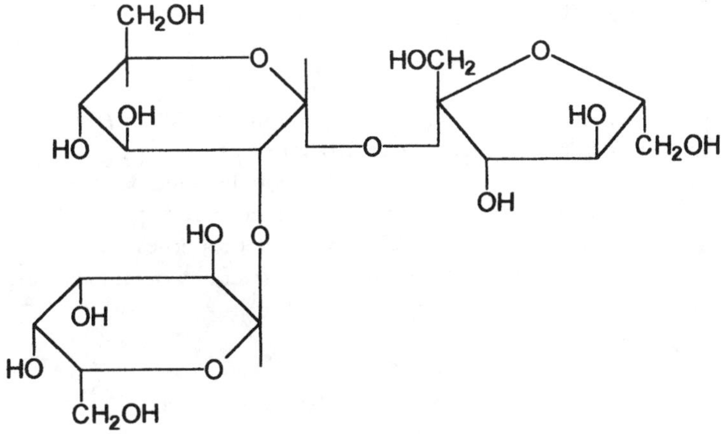

Figure 1a Umbelliferose

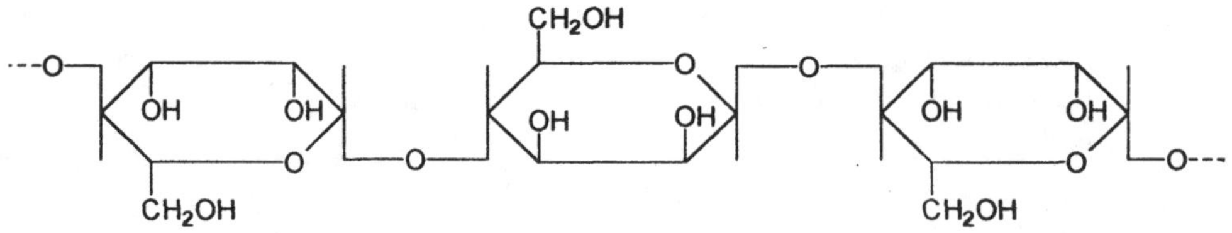

Figure 1b   The fragment of β-mannan

means of radioisotope techniques by Hopf and Kandler (1976). The biosynthesis of umbelliferose in vivo proceeds by the transfer of a galactosyl residue from an activated precursor to saccharose. During the synthesis of reserve cellulose in the endosperm large amounts of trehalose are also formed. While trehalose disappears almost completely during ripening, umbelliferose is accumulated continuously.

Monosaccharide composition of polysaccharide preparation from caraway is described by Hopf and Kandler (1977). The crude polysaccharide isolated from the fruit and seed coats consists mainly of pentoses: arabinose and xylose, hexoses: glucose together with mannose and galactose present in smaller amounts. Pentoses dominate in the crude polysaccharide fraction from the embryo, containing also traces of apiose, deoxysugar rhamnose and galacturonic acid, the later two present also in fruits and the seed coat. Additionally Bauer *et al.* (1988) identified deoxysugar fucose in polysaccharide fraction isolated from the seed and subjected to acid hydrolysis by means of trifluoroacetic acid.

Starch-free endosperm of *Carum* contains so called reserve cellulose deposited in cell walls. Fractionation and purification of this polysaccharide followed by acid hydrolysis, enzymatic breakdown and/or acetolysis, has led to its characterization β(1–4) mannan (Figure 1b). It is interesting to note that it has similar chemical structure as endosperm polysaccharide of taxonomically very different palm *Phoenix dactylifera* and *Phytelephas macrocarpa*.

3.3.2. Lipids of *Carum carvi* Seeds

Lipids constitute the group of substances possessing similar stuctural features to terpenes so they are often separated together from plant tissues.

Thus isolation procedure comprises an extraction by means of low polar organic solvent and subsequent distillation in vacuum. Terpenes are vaporised, unvolatile components separated according their specific chemical properties or molecular weights. Stepanenko *et al.* (1980) characterizes the lipid fraction obtained from Caucasian caraway in the following way [% of dry weight (d.wt.)]:

triglycerides	66
free fatty acids	5.1
steroids	0.4
hydrocarbons	0.2

chlorophyll   0.2
waxes   0.1
free alcohols   0.1

According to him, the acid components of triglycerides or occuring in a free form are both saturated and unsaturated in the range $C_{14:1}$–$C_{18:3}$ with $C_{18:2}$ acid predominating. It is worth to note that petroselinic acid $C_{18:1(6)}$, specific to *Umbellifereae* family is found in considerable amounts in both free (29%) and esterified (33.7%) form (Figure 2).

The other group of constituents isolated usually in lipid fraction are polyacetylenic compounds and polyenes. They are structurally and biogenetically related to lipid acids: their biological activity is still not precisely recognized. Acetylenes occur mainly in plant roots and in the case of caraway two of them are identified by Bohlmann *et al.* (1961). These are falcarinodione (Figure 3a), and falcarinolone (Figure 3b), extremely reactive low polar compounds, possessing two conjugated triple bonds and some oxygen-containing functional groups. They occur in fresh roots in small amount. Acetylenic compounds are reported also as constituents of all three plants used in Chinese medicine as "Fang Feng II" drug, composed of *Carum carvi*, *Ligusticum bractylobum and Seseli mairei* roots (Baba *et al.* 1989).

Additionally Kartnig and Moeckel (1973) report the occurrence of phytofluene ($=\psi,\psi'$-carotene) in *Carum* roots. Higher hydrocarbons of caraway seeds are characterized by means of gas chromatography combined with mass spectroscopy methods (GC-MS) as C_{27}–C_{31} alkanes with C_{29} and C_{31} predominating. Saturated fatty acids with 12 up to 26 carbon atoms are the acid components of waxes.

Total sterol content in glyceride oils, estimated by means of GC, is 0.2–0.7%. Studies on sterol composition of some glyceride oils from *Apiaceae* (cumin, caraway, fennel among them) were performed lately by Zlatanov and Ivanov (1995). β-Sitosterol and stigmasterol were identified as major (35–40%) components of the sterol mixture.

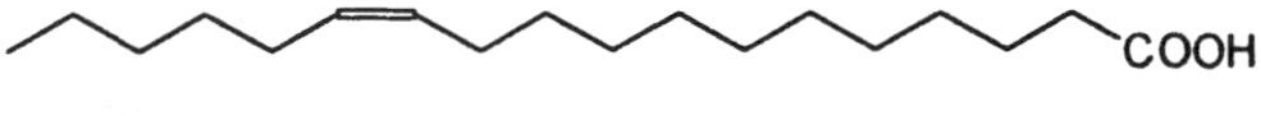

Figure 2 Petroselinic acid

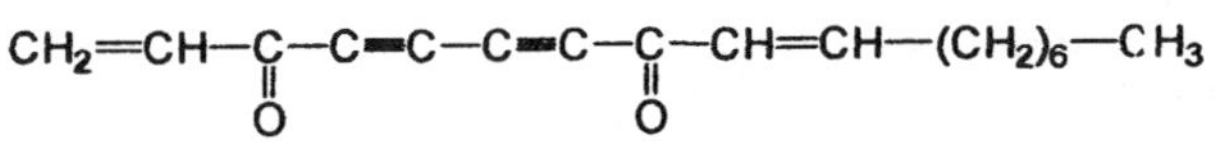

Figure 3a Falcarindione

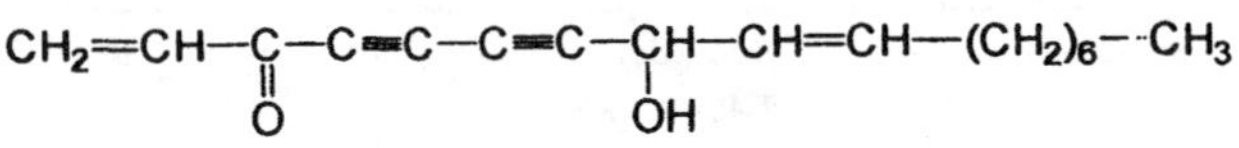

Figure 3b Falcarinolone

Its minor components were recognized as: cholesterol, brassicasterol, campesterol and Δ^7-campestrol, Δ^7-stigmasterol, Δ^5- and Δ^7-avenasterols.

3.3.3. Amino Acids and Peptides of *Carum*

It was commonly early recognized that caraway seeds are rich in peptide and protein substances. The composition of nitrogen-containing compounds in *Carum c.* was established more precisely in early eighties. So, free amino acids (FAA) and protein amino acids (PAA) of 28 species of medicinal herbs grown in Rumania, were qualitatively determined by Perseca *et al.* (1981). Free amino acids were estimated in homogenized tissues (leaves, stems, flowers, fruits) and water decoctions prepared thereof, by means of bi-dimensional thin layer chromatography (TLC). PAA composition – in acid hydrolysates of proteinic precipitates. The results of above studies are presented in Table 2 (column 2 and 3). Free alanine, phenylalanine, methionine, glutamic acid, serine and valine prevail in caraway seed decoctions. Proteic amino acid composition was found dependless on the nature of the organ studied.

Table 2   Amino acid composition in *Carum carvi* seed extracts and hydrolysates

1. Amino acid (a.a.)	2.. Free a.a. fraction (Perseca 1980)	3. Proteic a.a. fraction (Perseca 1980)	4. Proteid a.a. fraction (Brieskorn 1971)*
Alanine	++	+	+
Arginine	—	—	++
Aspartic acid	+	+	++
asparagine	+	+	—
cisteinic acid	+	—	—
GABA**	—	+	—
glycine	+	+	+
glutamic acid	++	++	+++
histidine	+	—	+
isoleucine	—	—	+
leucine	+	—	++
lysine	+	—	+
methionine	+	—	++++
ornitine	+	—	—
phenylalanine	++	—	+
proline	+	+	+
serine	++	+	+
threonine	+	+	+
tyrosine	+	+	+
valine	++	—	—

* Determined quantitatively for the following ranges in proteid fraction
  hydrolysates: (+) = 1.7–4.4%, (++) = 4.4–6.6%, (+++) = 8–12%,
  (++++) = 13.3%
** GABA = γ-aminobutyric acid

Proteid fraction of caraway fruits was studied by Brieskorn *et al.* (1972). It was found that *Carum* seeds, quite like anise ones, contain up to 3% of water-soluble, brown-coloured proteid with protein content ca 73%; the last one is composed of 16 different amino acids citied in Table 2, column 4. Molecular weight of caraway proteid is 30,000 D. The non-proteic fraction contains the following saturated fatty acids with general formula $C_nH_{2n}O_2$: arachinic (n = 20), behenic (n = 22) and palmitic, as well as unsaturated ones: oleic *cis*-$C_{17}H_{33}COOH$, linoleic $C_{17}H_{31}COOH$ and linolenic $C_{17}H_{29}COOH$. Acids were identified as methyl esters by means of GC methods. Alkaline (and reductive) hydrolyses of the proteid yield phenolic compounds and some derivatives of cinnamic acid, described with more details in subchapter 3.4.4.

Free glutamic acid content, determined for the first time by Gerhardt and Schulz (1985) by means of iodonitrotetrazolium chloride/diaphorase method, is 0.10–0.11% of dry weight.

Elemental composition of the protein bodies from caraway seeds is described in subchapter 3.4.5.

3.3.4. Abscisic Acid in *Carum*

Umbellifers are not easy to grow from seeds. The studies undertaken in order to find out the reason of such poor germination have revealed, among others, the occurrence of free endogenous abscisic acid (ABA) in *Umbelliferous* fruits, *Carum carvi* among them. The presence of this compound was substantiated by combined GC-MS and quantified by optical dispersion methods. Abscisic acid (syn. dormin, abscisin II) is known as an abscission-accelerating plant hormone and its structure is presented on Figure 4.

Natural ABA is the mixture of two stereoisomers *cis–trans* (Figure 4a) and *trans–trans* (Figure 4b), the first usually predominating with the ratio specific for a given plant. Caraway seeds contain 120 µg/kg of d.wt. ABA, with 76% predomination of (+) *cis–trans* isomer (Mendez 1978). It is recognized as medium-range content of ABA in comparison to other Umbellifers. A linear relationship between ABA content and dormancy degree in caraway seeds was found by Hradlik and Fiserova (1980).

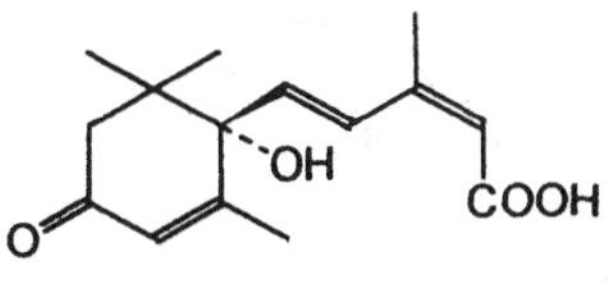

Figure 4a   (*cis–trans*) Abscisic acid

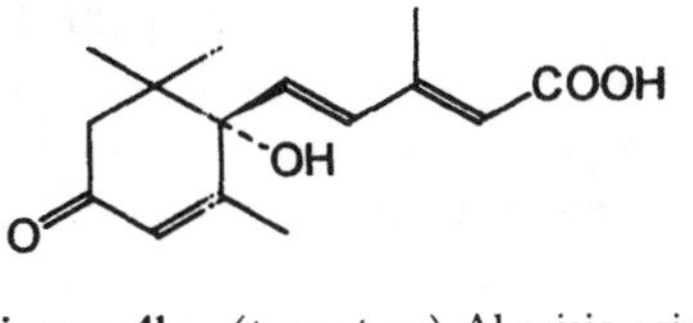

Figure 4b   (*trans–trans*) Abscisic acid

They observed also that cold stratification at 4°C for 20–40 days of pretreated seeds decreases ABA content, both endo- and exogenous.

3.3.5. Minor Miscellaneous Constituents of *Carum*

It happens frequently that minor constituents in plant extracts are discovered by chance in the course of a search directed towards specified group of natural compounds. That is the case free natural organic acids found in *Carum* seeds. So, Mendez (1978) states the isolation of p-hydroxybenzoic, vanillic and p-coumaric acids, ubiquitously present in plants (see also 3.4.4). Hopf and Kandler (1976) relate the identification of glyceric, apple and citric acids, present in water extracts obtained during biosynthetic experiments, Swain *et al.* (1985) estimate salicylic acid content. Sugar-derived alcohol–inositol is mentioned as well.

Matsuzawa and Kawai (1996) describe a simple and rapid method for quantitative determination of choline in foods by high performance liquid chromatography (HPLC) method. Choline is nitrogen-containing constituent of phospholipidic cellular walls. The hydrolysis of combined forms of choline in foods (caraway seeds included) performed by means of hydrochloric acid leads to choline chloride. The average concentration of so liberated choline in spices is within the range of 0.030–0.15% of d.wt.

3.4. SECONDARY METABOLITES OF *CARUM CARVI*

The following groups of secondary metabolites are described as constituents of caraway: terpenes, flavonoids, steroids, coumarins, tannins and phenolic substances. Secondary metabolites in contrast with primary compounds may be tissue specific.

3.4.1. Terpenes of *Carum carvi* Essential Oil

Terpenoid constituents identified so far in *Carum carvi* essential oil are presented on Figure 5. The biogenetic relationships between most of them, originating from a common precursor – geranyl pyrophosphate, is shown on the Figure 6.

The biosynthesis of monoterpenes proceeds from acetate units in the form of acetyl coenzyme A, initially along the same mevalonic acid pathway that gives rise to steroids. The key intermediate for entry to monoterpenes is neryl pyrophosphate, the geometric isomer of geranyl pyrophosphate. This pyrophosphate forms easily the reactive carbocation A, whose subsequent reactions of deprotonation or rearrangement to other carbocations (B or C) can be used to rationalize the formation of a wide range of acyclic, monocyclic or bicyclic hydrocarbons or oxygen-containing products.

Many physiologically active constituents identified in *Carum* seed are low molecular weight monoterpenes with oxygen-containing functional groups. Predominating oil constituent, S(+) carvone, formerly carvol, wide known also as D-carvone, results most probably from allylic oxidation of R(+) limonene with carveol intermediate (Glidewell 1991). Quantitative determination of such complicated mixture of structurally similar natural products as occurs in the essential oil, became possible due to a great progress in separation techniques made recently (thin layer chromatography – TLC, gas

Oxygenated terpenes:

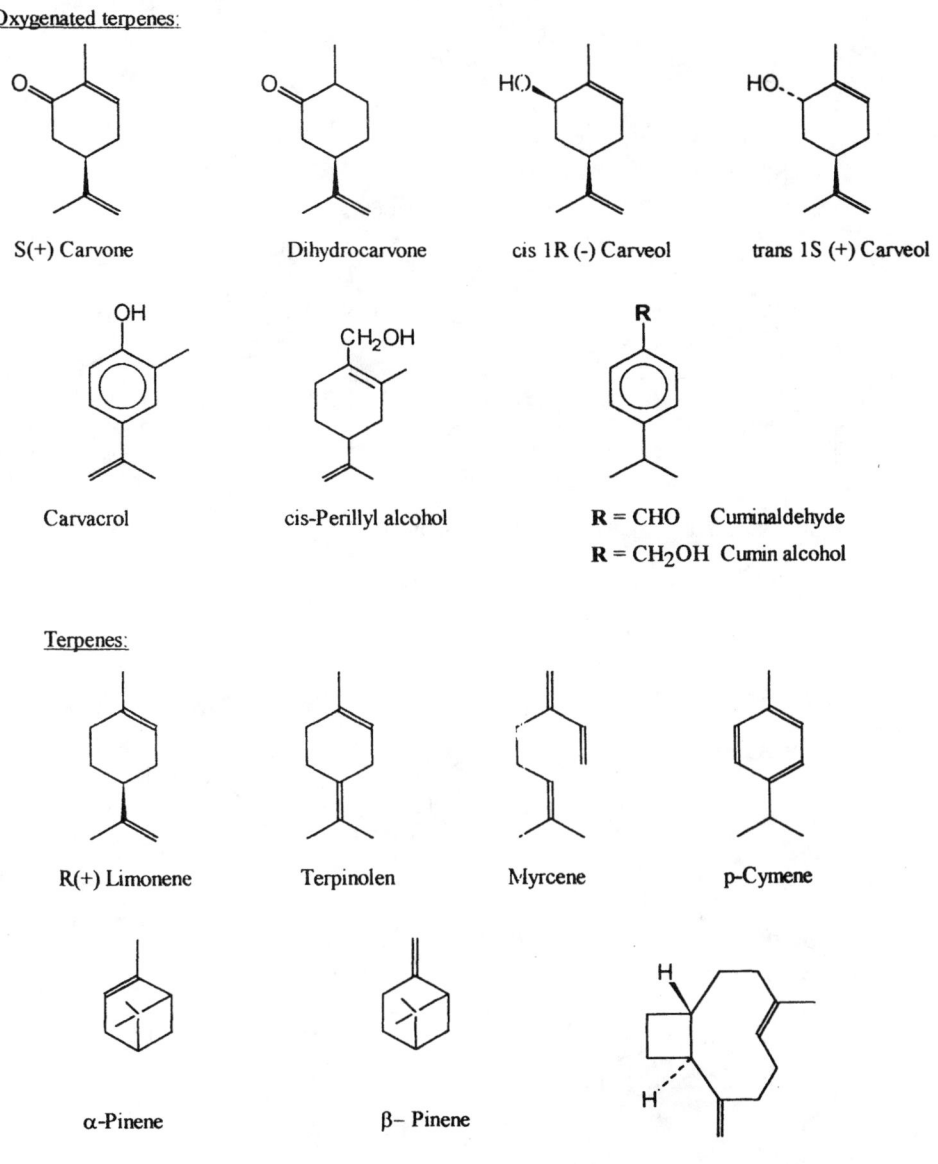

S(+) Carvone   Dihydrocarvone   cis 1R (-) Carveol   trans 1S (+) Carveol

Carvacrol   cis-Perillyl alcohol   R = CHO   Cuminaldehyde
                                            R = CH₂OH Cumin alcohol

Terpenes:

R(+) Limonene   Terpinolen   Myrcene   p-Cymene

α-Pinene   β- Pinene

β-Caryophyllene

Figure 5   Main constituents of *Carum carvi* essential oil

chromatography – GC, high performance liquid chromatography – HPLC on reversed phase and chiral sorbents). Especially this last technique – chiral HPLC together with the analysis of mass spectra made possible the identification of all stereoisomers produced by *Carum carvi*.

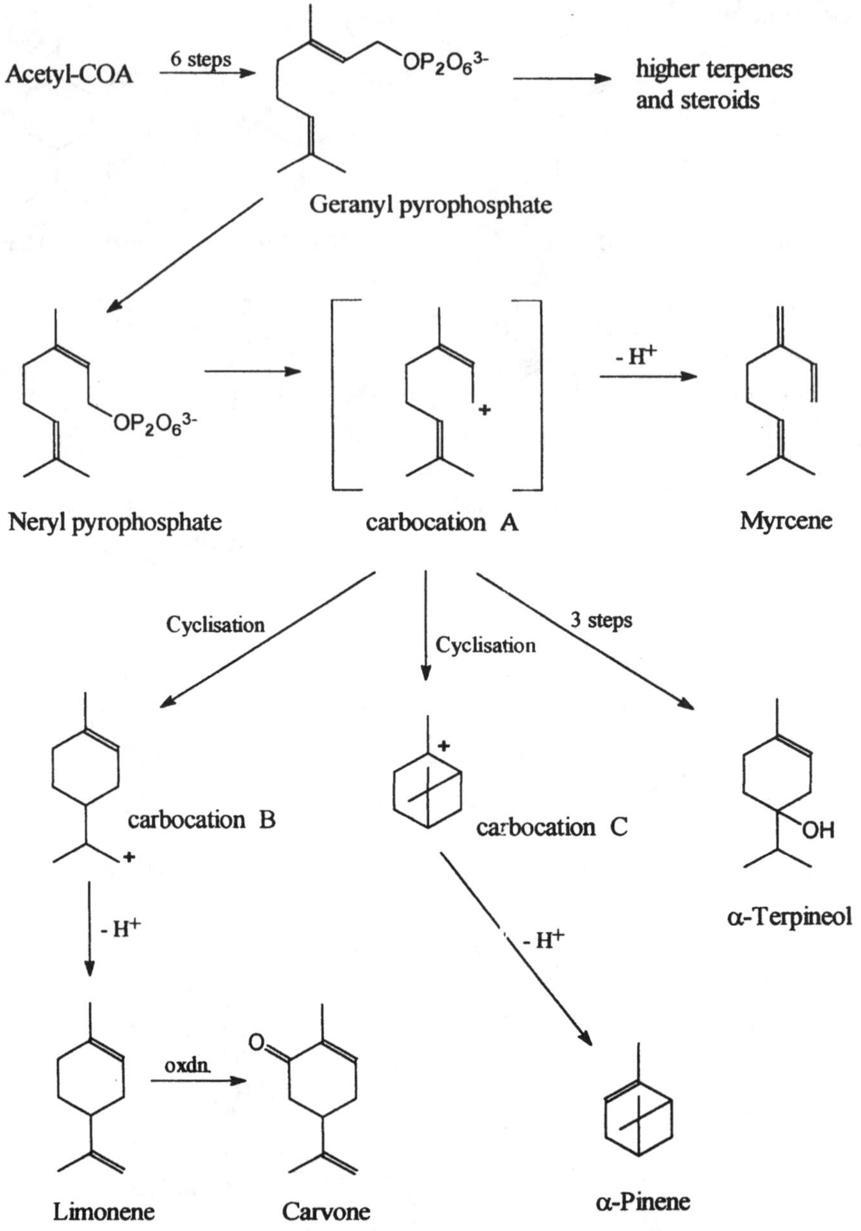

Figure 6 Biogenetic relationship of *Carum* terpenoic consistuents

Monoterpene R(+) limonene accompanies carvone in caraway seed volatile oil, with the ratio of both main components varying from 3:2 up to 3:1, depending on variety of plant and storage conditions. The content of other minor and trace substances in the oil may also vary within broad limits as shown in Table 3. Data presented in the first column are mean values determined for samples of caraway cultivated in mid-European countries while data in the second are estimated for caraway of Egyptian origin. The big discrepancies observed may result also from different isolation procedures used by both authors. El Wakeil *et al.* (1986b) observed the increasing of carvone content in the mixture during 3 first months of oil storage.

Carvone is α–β unsaturated ketone and this functional group may be essential for its biological activity as detoxifying enzyme activity inducer (Zheng Guo Quiang *et al.* 1992). This group may react with SH function of GSH or GST in a manner of 1,4-nucleophilic addition following Michael reaction mechanism. This is consistent with the hypothesis that many anticarcinogenic enzyme inducers are Michael reaction acceptors with electron-withdrawing groups conjugated with olefinic bonds. Electrowithdrawing group potency parallels the inducer efficiency in detoxyfication process.

S(+) Carvone, as dominating constituent in caraway essential oil is considered a diagnostic substance, for example for TLC analysis of mixtures of fluid extracts of caraway, anis, fennel, lovage, juniper and melissa.

Minor constituents – carvacrol, cumin alcohol and cuminaldehyde found in non-volatile part of caraway essential oil belong to phenolic substances. Recent search for constituents responsible for antioxidant properties of *Carum* has led, among others, to carvacrol (Lagouri and Boskau 1995).

Table 3 The composition of caraway essential oil

1. Constituent	2. [% of d.wt.] (Pushman 1992)	3. [% of d.wt.] (El Wakeil 1986a)
Total essential oil	5.36	0.99
S(+) carvone	50.46	80.17
R(+) limonene	47.66	9.75
Dihydrocarveols	0.56	0.04
Dihydrocarvones	0.18	0.70
myrcen	0.35	0.06
carvylacetate	0.16	undtmd
carveols (*cis* & *trans*)	undtmd	0.24
pinenes (α+β)	undtmd	0.06
terpinolene	undtmd	0.20
β-caryophylene	undtmd	0.11
p-cymene	undtmd	0.06
cumin aldehyde	undtmd	0.08
cis-perillyl alcohol	undtmd	0.14
cuminalcohol	undtmd	0.02
unidentified compounds	less than 1%	8.17

undtmd = undetermined by authors

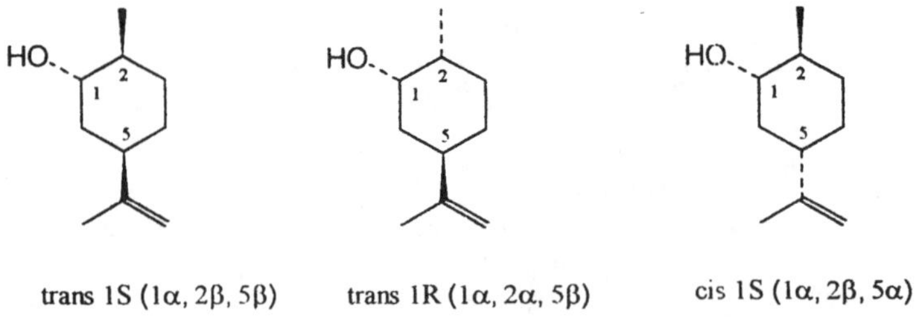

trans 1S (1α, 2β, 5β)   trans 1R (1α, 2α, 5β)   cis 1S (1α, 2β, 5α)

Figure 7 Isomeric dihydrocarveols

Dihydroderivatives of main terpenes – dihydrocarvone and dihydrocarveol may occur as mixtures of stereoisomers due to appearing of further chiral carbon centers in the molecule. Their resolution became possible after application of HPLC on permethylated β-cyclodextrin. Stereochemical representation of possible isomers as well as both isomeric carveols are shown on the Figure 7. Terpenoic constituents of *Carum*, beside appreciated biological activity, are of interest for organic chemists as natural building blocks used in enantioselective syntheses of natural compounds.

3.4.2. Flavonoids of *Carum*

Flavonoids (flavonoid glycosides) are the other important group of *Carum* secondary metabolites relatively early studied and recognized due to diversified biological activity this group of compounds usually exerts. Flavonoid pattern in the fruits of *Umbellifereae* plants was surveyed by Harborne and Williams (1972). Fruit as well as leaf tissue can be usefully employed to provide chemical characters for taxonomic purposes. For flavonoid surveys only small samples of seeds are required (usually 5–10 fruits); flavonoids are very evenly distributed throughout the family, more than other chemicals. According to Harborne and Williams (1972) there is a major division in the subfamily *Apioideae* between tribes producing flavonoid alcohols (flavonols) and those synthesizing more oxidized flavones. *Carum carvi* (*Apiaceae*) belongs to the flavonolic group together with *Coriandreae, Smyrniceae* and others.

Seed flavonoids occur in the form of 3-O-glycosides; their isolation, identification and quantitative determination was performed by Kunzemann and Herrmann (1977) by means of chromatography on cellulose columns and HPLC methods. The following compounds were obtained crystalline from caraway seed methanolic extract:

quercitin 3-glucuronide,
isoquercitrine = quercetin-3-O-β-glucopyranoside,
quercetin-3-O-caffeylglucoside,
kaempferol-3-glucoside.

Structures of glycoside aglycones, quercetin and kaempferol, as well as glycoside isoquercitrine are presented on Figure 8. All four compounds occur also in *Carum* leaves, which contain also in addition isorhamnetin glycosides in lower concentration.

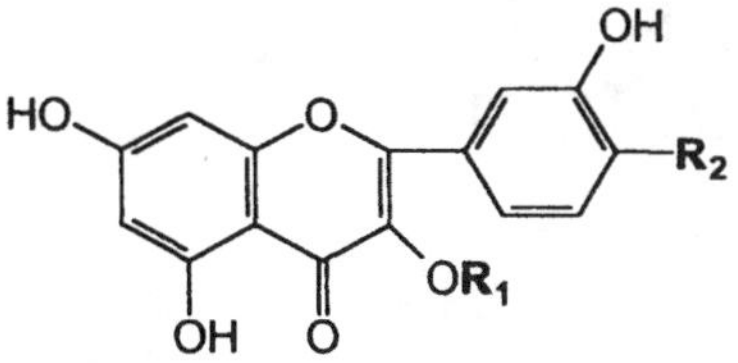

R_1	R_2	
H	OH	Quercetin
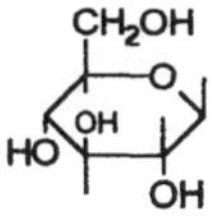 CH₂OH sugar	OH	Isoquercitrine
H	H	Kaempferol

Figure 8 Examples of Carum flavonoid components

Predominating constituent, isoquercitrine, occurs in caraway seed in the amount 80 mg per kg of d.wt., while others are found in the quantity 2 up to 10 times lower.

The main feature of biological activity they exert is their antioxidative properties. It is recognized also that they may activate enzymes detoxifying carcinogenic substances and metabolites in cells.

3.4.3. Coumarins of *Carum*

Coumarins, and especially furocoumarins, constitute also the group of natural compounds of wide spectrum of biological activity. They are antibacterial, potent photosensitizers when activated by near UV-light, they intercalate readily into DNA and form light-induced mono- and di-adducts with pyrimidine bases, thus they are phototoxic, mutagenic and photocarcinogenic. Moreover, they exhibit strong seed germination inhibiting action.

There were some reasons at the very beginning of the search directed towards coumarinic group of constituents in *Umbellifereae* family:

1. – the need for new natural sources of coumarins after the discovery of their possible utilisation in psoriasis treatment and in sunscreen lotions and preparations,
2. – dietary concern after the constatation of toxic effects of coumarins in the seeds of *Umbellifereae* plants and oils obtained from them in the view of their big consumption as spices and flavorings,
3. – poor germination of *Umbellifereae* seeds.

The occurrence of coumarins in caraway seeds was first described by Nielsen (1970), who mentioned identification of umbelliferone, coumarin and scopoletin. (Figure 9a). Kaminski and coworkers (1978) found 3 to 5 coumarinic compounds in ethyl acetate extracts of various tissues of *Carum*, but the only compounds identified on

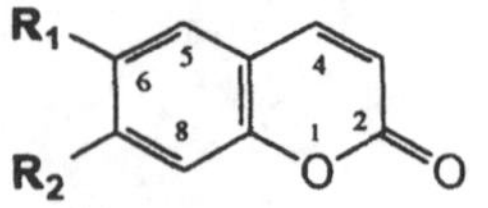

R$_1$	R$_2$	
H	H	Coumarin
OCH$_3$	OH	Scopoletin
H	OH	Umbelliferone

Figure 9a

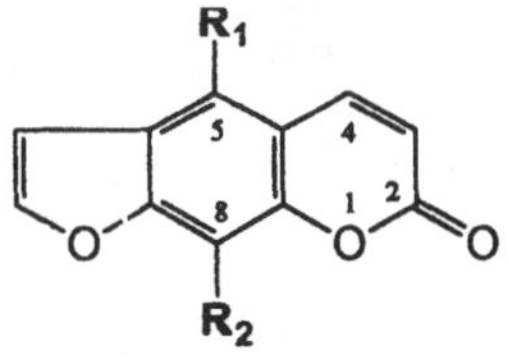

R$_1$	R$_2$	
OCH$_3$	H	5-MOP
H	OCH$_3$	8-MOP

Figure 9b   Coumarinic consistuents of *Carum carvi* seed

the basis of colour reactions and chromatographic behavior were scopoletin and coumarin. Only when ultrasensitive bioassay was applied by Ceska and coworkers (1987), two others furocoumarins: 8-methoxypsoralen (8-MOP) and 5-methoxypsoralen (5-MOP) were detected, identified and quantitatively determined (Figure 9b). The furocoumarin content in caraway seeds is extremely low: both occur approximately 0.005 µg/g of d.wt. while bioassay detection limit is 0.001 µg/g.

3.4.4. Phenolic Compounds

Well-known and well-documented stabilizing effect of spices (caraway included) on food, especially meat products was from the very beginning attributed to their constituents with phenolic functional groups. The above effect may derive from antimicrobial or antioxidant properties of active substances.

There are many phenolic substances found so far in caraway seed. Carvacrol is mentioned in the subchapter 3.4.1. The occurrence of free phenolic acids, substantiated by some authors, is citied in subchapter 3.3.5. The results of some more advanced studies have revealed however that the last group of compounds can be regarded as artifacts liberated from natural derivatives by hydrolysis or during isolation procedures.

The following compounds containing phenolic fragments are so far identified in *Carum* fruit:

– flavonoids (Harborne and Williams 1972) and others (see also **3.4.2.**),
– glycosides (Dirks and Hermann 1984),
– derivatives of quinic acid (Dirks and Hermann 1984),
– proteids (Brieskorn *et al.* 1972),
– tannins (Zia-ur Rehman *et al.* 1993),

So, Dirks and Herrmann (1984) has identified the 4-(β-D-glucopyranosyloxy)benzoic acid (Figure 10) in water extracts obtained from ground dried seeds and green parts, in very mild conditions and estimated its content as 40 ppm. This compound is a natural glycoside of hydroxybenzoic acid mentioned already in subchapter 3.3.5 together with p-coumaric acid, both most probably being artifacts. Four derivatives of cyclohexanecarboxylic (quinic) acid (Figure 11a) and p-coumaric acid (Figure 11b) are identified as well. These are: 3-, 4-, and 5-caffeoylquinic acids (Figure 11c) and 3-o-p-coumaroylquinic acid (Figure 11d) found in the amount of 430, 310, 125 and 50 ppm in caraway and in other spices. Antimicrobial tests with the use of the above phenols indicate that they are not responsible for stabilising effect of spices.

The identification of phenol-containing proteid by Briescorn *et al.* (1972) was mentioned already in the subchapter 3.3.3. Splitting off the protein under reductive conditions liberates 3,4-dihydroxycinnamic (caffeic) acid (Table 4). Detailed studies of acidic, alcaline, enzymatic and reductive cleavage products lead to the conclusion that caffeic acid is bound to proteid as a result of nucleophilic 1,4-addition of free proteic amino group to o-quinoic intermediate with the formation of C–N bond between nitrogen and C-6 atom of the benzene ring (Figure 12).

The most comprehensive spectrum of phenolic acids obtained after enzymatic release from their naturally occurring derivatives in most common spices and then identified and determined by means of GC-MS methods and HPLC is supplied by Schulz and Herrmann (1980). Derivatives of hydroxybenzoic and hydroxycinnamic acids identified in caraway seeds are presented together with relevant contents in Table 4.

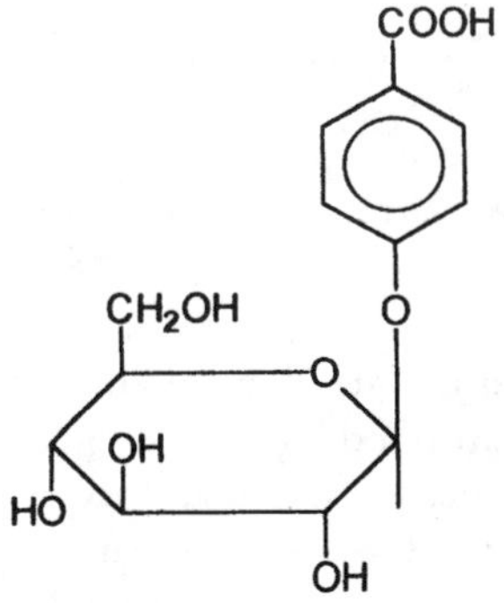

Figure 10 4-(β-D-glucopyranosyloxy) benzoic acid

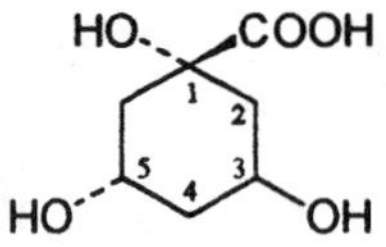

Figure 11a   Quinic acid

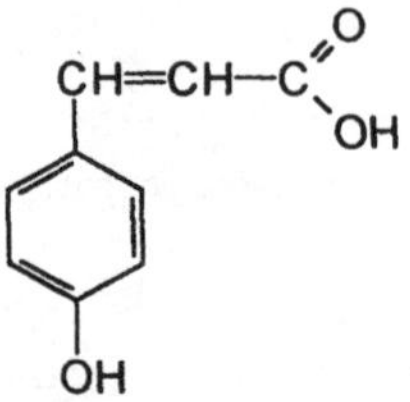

Figure 11b   p-coumaric acid

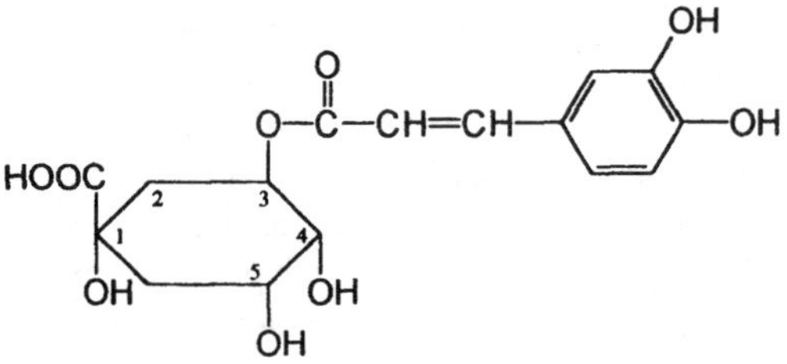

Figure 11c   3-O-caffeoylquinic acid

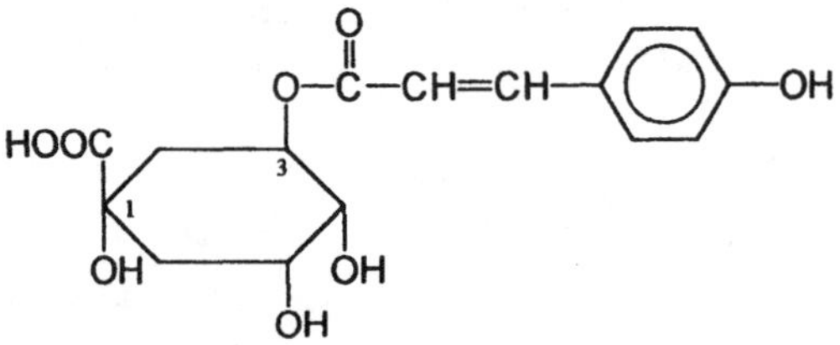

Figure 11d   3-O-coumaroylquinic acid

Redox behavior of spices can be quantitatively estimated by means of potentiometric methods as described by Gerhardt and Boehm (1980). They found that redox potential of food (spice) samples suspended in water and expressed as R_H value dependent on pH or $R_{H'}$ value calculated for pH$=7$ is direct inversely proportional to total phenol content. Spices can be divided into two groups according to their $R_{H'}$ values; those with $R_{H'}$ ranging from 0 to 9 exhibit strong reductive properties while those with highest

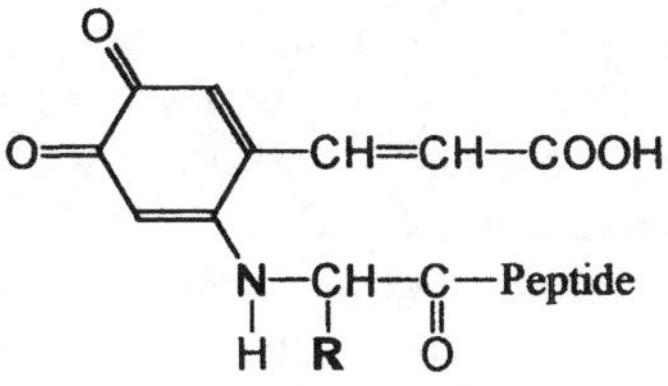

Figure 12 Fragment of phenol-containing proteid molecule

Table 4 Derivatives of hydroxybenzoic and hydroxycinnamic acids found in caraway seed according Schultz and Herrmann (1980)

Derivatives of aromatic acids:

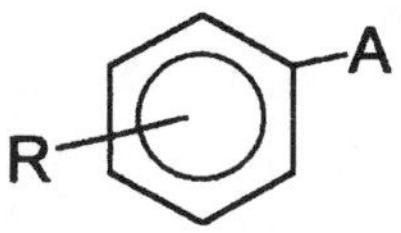

A = COOH; hydroxybenzoic			A = CH = CH-COOH; hydroxycinnamic		
name	*substituent*	*content* [ppm]	*name*	*substituent*	*content* [ppm]
Salicylic	2-OH	9	p-coumarinic	4-OH	86
p-hydroxybenzoic	4-OH	70	caffeic	3,4-di-OH	2840
Gentisinic	2,5-di-OH	10	ferulic	3-OCH$_3$, 4-OH	170
Protocatechic	3,4-di-OH	64	sinapic	3,5-di-OCH$_3$ 4-OH	10
Vanillic	3-OCH$_3$, 4-OH	25			
Syringic	3,5-di-OCH$_3$, 4-OH	40			

$R_{H'}$ values (34–42) are strong oxidants. Caraway ($R_{H'} = 3.33$) has strong reductive activity, together with aniseed ($R_{H'} = 3.18$), capsicum ($R_{H'} = 2.16$), allspice ($R_{H'} = 3.26$) and majoram ($R_{H'} = 2.67$). Total phenol content in caraway seed is established as 0.52% of d.wt.; this result is consistent with value obtained undependently by colorimetric methods. So, well-known stabilising effect of some spices on food, especially on meat products, has gained support and explanation in their relatively high content of phenolic substances.

3.4.5. Elemental Composition of *Carum*

Caraway fruit is characterized by relatively high ash content (Table 1), resulting from mineralization of compounds being its elemental reserves. A great part of these mineral reserves is located inside the protein bodies. According to Spitzer *et al.* (1982), who studied the elemental composition of globoid crystals in the caraway endosperm by means of energy-dispersive X-ray analysis, they contain potassium, calcium, phosphorus and magnesium, the two first ones present usually in higher amounts. The proteinaceus matrix contains sulfur and phosphorus, regardless of the inclusion present in the protein body. More than 90% of calcium is located in the endosperm and pericarp portions of the mericarp. The chemical composition of Ca-rich crystal inclusions in the seed protein bodies is identified as calcium oxalate. Oxalate content in caraway, as well as in other spices estimated as ranging 0.01–0.02% of d.wt. is considered relatively high. According to Ramasastri (1983) caraway seed is also rich in free oxalic acid (0.06–0.09% of d.wt.), so condiments should not be neglected in lithiasis risk assessment studies.

Last two decades are characterized by a great development in analytical methods directed towards estimation of elemental composition of natural materials, i.e. atomic absorption spectrometry (AAS), potentiometric stripping analysis (PSA), voltamperometry and instrumental neutron activation analysis (INAA). Due to that many locally produced condiments and spices were lately carefully inspected regarding their trace elemental composition, essential as well as toxic elements or impurities. So, Khan et al. (1985) evaluated trace elements of spices grown locally in Pakistan or imported. He found that caraway is rich in all trace metals: iron, copper, nickel, zink, manganese, and cobalt, especially abundant in first three of them. Contents of trace elements in Carum originating from various sources determined by different authors are gathered in Table 5. Additionally Peters (1986) estimated selenium content in Carum as ranging 0.1–0.38 μg/100 g of d.wt. It is worth to note only small discrepancies in elemental levels established by two independent methods (AAS and INAA). Regarding high content of iron, copper and zinc, caraway may be recommended diet allowance important for adequate nutriture.

Monitoring of toxic metals content in foods is now more and more frequently performed with the use of methods mentioned above in the view of growing environmental pollution. So lead, cadmium and thalium contents of Fructus carvi from The Netherland, Germany and Poland estimated by AAS are respectively the following [ppm]: 0.01–0.13; 0.05–0.12 and less than 0.01 (Laik 1983) and is strongly site-dependent. Weekly tolerable intake limit set by FAO/WHO Expert Committee is lead – 3 mg, cadmium – 0.315–0.330 mg per person.

Table 5 Content of trace elements in Carum according to Khan (1985) and Saffar (1989)

Element	Determined by Khan (1985) AAS method [μg/g]	Average literature value [μg/g]	Determined by Saffar (1989) PSA method [μg/g]
Iron	286.0		300.1
Copper	13.8	6.8	12.9
Zinc	47.0	22.9	49.5
Manganese	33.8		38.8
Chromium	15.5	3.3	—
Nickel	7.0	0.16–2.75	2.75
Cobalt	1.0	0.5	—

Increasing environmental pollution and inadequate agricultural treatment of plants and/or crops results sometimes in occurrence of toxic impurities in them. The following substances were, among others, found and monitored in *Carum* seed: pesticides – dimethoate, heptachlor, aldrin, dieldrin, DDE, o,p- and p,p-DDT, α-, β-, γ-H-C-H, HCB and B-58, fluorides, and nitropolycyclic aromatic hydrocarbons (PAH). It is worth to note that caraway exhibited relatively high residue of dimethoate and B-58, most probably caused by the presence of reactive functional group in the molecule. Respective references are available from the author on request.

REFERENCES

Abou-Zied, E.N. (1974) Increase in volatile oil and chemical composition in the seeds of caraway and fennel plants induced by succinic acid 2,2-dimethylhydrazide. *Biol. Plant.*, **16**, 123–126.

A.O.A.C. (1975) *Association of Official Analytical Chemists, Official Methods of Analysis*, 12th ed., The Association, Washington, D.C.

Atal, C.K. and Sood, N.M. (1967) Study of Indian caraway and its substituents. *The Indian J. of Pharmacy*, **29**, 42–44.

Baba, K., Yoneda, Y., Kozawa, M., Fujita, E., Wang, N. and Yuan, C. (1989) Studies on Chinese traditional medicine "Fang Feng" II. *Shoyakugaku Zasshi*, **43**, 216. *CA* 112, 240358x.

Bauer, F., Vali, S. and Stachelberger, H. (1988) Composition and content of carbohydrates in spices. *Chem. Mikrobiol. Technol. Lebensm.*, **11**, 181–191.

Bohlmann, F., Arndt, C., Bornowski, H. and Kleine, K.-M. (1961) Über polyine aus der Familie der Umbelliferen. *Chem. Ber.*, 954–960.

Brieskorn, C.H., Hagen, P. and Mosandl, A. (1972) O-Diphenolproteide aus den Früchten des Anis und des Kümmels. *Z. Lebensm.-Unters. Forsch.*, **148**, 83–89.

Ceska, O., Chaudhary, S.K., Warrington, P.J. and Ashwood-Smith, M.J. (1987) Photoactive furocoumarins in fruits of Umbellifers. *Phytochemistry*, **26**, 165–169.

Dirks, U. and Herrmann, K. (1984) HPLC of hydroxycinnamoyl-quinic acids and 4-(β-D-glucopyranosyloxy)benzoic acid. *Z. Lebensm.-Unters. Forsch.*, **179**, 12–16.

El-Wakeil, F., Khairy, M.S. Morsi, Farag, R.S., Shihata, A.A. and Badei, A.Z.N.A. (1986) Biochemical studies on the essential oils of some fruits of Umbellifereae family. *Seifen-Oele-Fette-Wachse*, **112**, 77–80.

El-Wakeil, F., Khairy, M. Morsi, Farag, R.S., Shihata, A.A. and Badei, A.Z.N.A. (1986) Effects of various storage conditions on the quality of some spice essential oils. *Seifen-Oele-Fette-Wachse*, **112**, 348–353.

Gerhardt, U., and Boehm, Th. (1980) Das Redox-Verhalten von Gewurzen in Fleischerzeugnissen. *Fleischwirtsch.*, **60**, 1523–1526.

Gerhardt, U. and Schulz, W. (1985) The presence of free glutamic acid in foods. *Fleischwirtsch.*, **65**, 1483–1486.

Glidewell, C. (1991) Monoterpenes, an easily accessible but neglected class of natural products. *J. Chem. Educ.*, **68**, 267–269.

Harborne, J.B. and Williams, C.A. (1972) Flavonoid patterns in the fruits of the Umbellifereae plants. *Phytochemistry*, **11**, 1741–1750.

Hopf, H. and Kandler, O. (1976) Physiologie der Umbelliferose. *Biochem. Physiol. Pflanzen*, **169**, 5–36.

Hopf, H. and Kandler, O. (1977) Characterization of the "Reserve cellulose" of the endosperm of *Carum carvi* as a β(1–4) Mannan. *Phytochemistry*, **16**, 1715–1717.

Hradlik, J. and Fiserova, H. (1980) Role of abscisic acid in dormancy of caraway seeds. *Acta Univ. Agric. Fac. Agron. (Brno)*, **28**, 39–64.

Kaminski, B., Glowniak, K., Majewska, A., Petkowicz, J. and Szaniawska-Delundej, D. (1978) Search for coumarin compounds in fruits and seeds. *Farm. Pol*, **34**, 25–28.

Kartnig, T. and Moeckel, H. (1973) Lipid components from roots of *Carum carvi* and *Anethum graveolens. Sci. Pharm.*, **41**, 102–105.

Khan, H.H., Najma Bibi, G., and Mohiuddin Zia (1985) Trace metal contents of common spices. *Pakistan J. Sci. Ind. Res.*, **28**, 234–237.

Kunzemann, J. and Herrmann, K. (1977) Isolation and identification of flavon(ol)-O-glycosides in *Carum carvi* and *Foenicum vulgare. Forsch Z. Lebensm-Unters.*, **164**, 194–200.

Lagouri, V. and Boskau, D. (1995) Screening for antioxidant activity of essential oils obtained from spices. *Dev. Food Sci.*, **37A**, 869–879.

Laik, A.S. (1983) Determination of pesticide residues and other critical impurities – (Toxic trace metals) – in medicinal plants. *Pharm. Ind.*, **45**,1294–1295.

Matsuzawa, M. and Kawai, H. (1996) Determination of choline in spices by means of HPLC with an electrochemical detector. *Shokuhin Eiseigaku Zasshi*, **37**, 72–76. *CA* **125**, 84954z.

Mendez, J. (1978) Endogenous abscisic acid in Umbelliferous fruits. *Z. Pflanzenphysiol.*, **86**, 61–64.

Nathnagel and Rossbach (1883) *Farmakologija* 4th ed., Wydawnictwo Gazety Lekarskiej, Warszawa, p. 485.

Nielsen, B.E. (1970) *Coumarins and Umbelliferous Plants.* The Royal Dannish School of Pharmacy, Copenhagen.

Perseca, T., Girmacea-Sas, V., Cismas, V. and Lacan, E. (1981) The free and proteic amino acids in the homogenized tissues and in the decotion products of some medicinal herbs. *Memorille Sect. Stiint. Acad. Rep. Soc. Rom.*, **4**, 179–184.

Peters, H.J. (1986) Contents of some trace elements in foods purchased or produced in the area of Leipzig. *Spurenelem. Symp. 5th* Ed. Anke, M. Univ. Jena.

Pushmann, G., Stephani, F. and Fritz, D. (1992) Investigation on the variability of *Carum carvi. Gartenbauwissenschaft*, **57**, 275–277.

Ramasastri, B.V. (1983) Calcium, iron and oxalate content of some condiments and Spices. *Plant Foods Hum. Nutr. (India)*, **33**, 11–15.

Saffar, A. (1989) Concentration of selected heavy metals in spices, dry fruits and plant nuts. *Plant Foods Hum. Nutr. (India)*, **39**, 279–186.

Schulz, J.M. and Herrmann, K. (1980) Über das Vorkommen von Hydroxibenzoesauren and Hydroxizimtsauren. *Z. Lebensm-Unters. Forsch.*, **171**, 193–199.

Spitzer, E. and Lott, J.N.A. (1982) Protein bodies in Umbelliferous seeds. II. Elemental Composition. *Can. J. Bot.*, **60**, 1392–1398.

Stepanenko, G.A., Gusakova, S.D. and Umarov, A.U. (1980) Lipids of *Carum carvi* and *Foenicum vulgare* seeds. *Khim. Priror. Soedn.*, 827–828.

Swain, A.R., Dutton, S.P. and Treswell, A. (1985) Salicylates in foods. *J. Am. Diet. Assoc.*, **85**, 950–960.

Zheng Gou Qiang, Kenney, P.M. and Luke, K.T. (1992) Anethofuran, carvone and limonene; potential cancer preventive agents from dill weed oil and caraway oil. *Planta Med.*, **58**, 338–341.

Zia-ur-Rehman, Chaudhry, M.I. and Malik, M.A. (1993) Tannin content of some locally produced spices. *Sci. Int. (Lahore)*, **5**, 63–65.

Zlatanov, M. and Ivanov, S.A. (1995) Studies on sterol composition of some glyceride oils from Apiaceae. *Fett. Wiss. Technol.*, **97**, 381–383.

4. THE ROLE OF LIGHT AND TEMPERATURE IN THE GROWTH, DEVELOPMENT AND ACTIVE AGENT ACCUMULATION IN CARAWAY AND RELATED SPECIES

SEIJA HÄLVÄ

Laboratories for Natural Products, and Aromatic and Medicinal Plants, University of Massachusetts, Amherst, MA, USA

4.1. BACKGROUND

Plant growth and development are affected by several ecological factors and these same factors will influence the production of essential oils as well as other active agents in herb plants. The effect of external conditions on chemical composition of plants has been recognized for more than 80 years. The environment affects the essential oil concentration directly through metabolic processes and indirectly through plant growth. Metabolic changes affect the amounts of substances synthesized, and plant growth alters dry matter production and the proportion of different plant organs. Certain conditions may favor either the production of roots, above ground vegetative organs, or flowers and seeds (Bernáth 1986, Bernáth and Tétényi 1978, Clark and Menary 1979, Flück 1955, Savchuk 1976).

The concentration of active agents in plants results from their continuous formation and breakdown during plant growth. The total accumulation depends upon the genetic composition of the plant and it varies between genera and species. Within a species itself, altering the environmental and genetic composition will produce different quantities and types of essential oils (Bernáth and Tétényi 1978, Croteau 1986, Franz *et al.* 1975, Franz *et al.* 1984, Tétényi 1986).

The main ecological factors include light, temperature, wind, water, soil, and nutrients. Also other organic or non-organic factors such as pesticides, radioactive rain and other stress factors affect the growth, development, and active agent accumulation. Within the environment, light and temperature have a major role in plant growth and development, and in the production of essential oils.

Relatively few studies have been done on the effects of specific environmental factors on caraway (*Carum carvi* L.). Biennial caraway, native to humid hayfields and hillside pastures of Middle Europe, prefers lots of sunshine and relatively little heat (Sváb 1992). Annual caraway, instead, requires more heat, and the plant is grown mainly in the Mediterranean countries (Putievsky 1978). This review will discuss the role of light and temperature on the growth and essential oil of caraway and, because of the very few studies on the species, also on some other essential oil producing plants. Essential oil yields of caraway vary greatly between the years mainly due to varying weather conditions. Apart from dill (*Anethum graveolens* L.), another *Apiaceae*, only peppermint (*Mentha x piperita* L.) and chamomile (*Matricaria recutita* L.) are more thoroughly studied.

4.2. THE ROLE OF LIGHT

4.2.1. Effect of Light Intensity

High fruit yield of caraway requires plenty of sunshine especially in the first year of growth and also during the flowering stage. Low light level will delay and decrease the fruit production (Smid and Bouwmeester 1993, Bouwmeester *et al.* 1995b). Biennial caraway is commonly seeded with a covercrop which is harvested before the caraway will be too shaded and therefore etiolated. The taproot should reach the diameter of at least 8 mm in order to the plant to flower in the following summer (Toxopeus and Lubberts 1994).

Increasing light intensity up to a saturation point causes an increase in the net carbon assimilation rate, and this, in turn, stimulates the vegetative growth and dry matter production in many plants. Increases both in herb yield and oil concentration in dill with increasing light intensity imply a direct connection between photosynthesis and oil accumulation (Hälvä 1993). On the other hand, low light intensity will produce plants with long internodes, thin and light colored leaves, and a smaller leaf area due to the lack of assimilates (Evans 1973, Firmage 1981, McLaren and Smith 1978, Morgan 1981, Morgan and Smith 1976, 1981).

Light intensity and essential oil formation are closely connected, since the precursors produced during the photosynthesis form the basic units necessary for the synthesis of essential oils (Clark and Menary 1980a, Lincoln and Langenheim 1978, Saleh 1973). According to Bouwmeester *et al.* (1995a), the quantity and quality of the essential oil in caraway is determined by the assimilate availability during the early stages of seed production. Low light intensity due to shading during the early stages of seed development will decrease oil accumulation and the proportion of carvone in the oil. Later in the seed filling, the shading will mainly decrease the size of the seeds. Seed weight and the absolute amount of essential oil per seed correlate positively for both the annual and biennial caraway.

The importance of carbohydrate assimilates is obvious in the case of caraway. Commonly reported lower essential oil content in the fruits of annual caraway (1–3%) compared to the fruits of biennial caraway (3–5%) in the field have been explained by genetic and climatic conditions. Bouwmeester and Kuijpers (1993), however, showed that the reported difference is caused by the lack of assimilates rather than genetic composition. The lack of assimilates results either from weather conditions or from unfavorable balance between the availability of assimilates for fruit growth and the number of fruits. Surprisingly, contrary to experiences in the field, annual and biennial caraway produced similar percentages of oil (4.5%) when grown under similar greenhouse conditions.

Global radiation, precipitation, and wind during ripening have been reported to affect the fruit yield and essential oil content in caraway. The negative effect of high precipitation and strong wind may also cause insufficient pollination and abortion of fruit, and therefore lower yields. Essential oil content in caraway correlates positively with the cumulative global radiation during seed filling and thus with the high rate of photosynthesis. The mechanism through which the light affects the yield still remains to unclear (Toxopeus and Bouwmeester 1992). The low carvone content in annual caraway can be increased by plant breeding. Crosses between a biennial caraway and an annual caraway have clearly increased the carvone content in the annual caraway (Toxopeus *et al.* 1995).

The interaction of light intensity and photoperiod significantly modifies plant growth. The effect of light intensity is greater under long day conditions than under short day conditions. Accordingly, the growth of peppermint, a long day plant, will increase under a combination of long day conditions and high light intensity (Clark and Menary 1980a). However, short day conditions with a high light intensity may result in as much growth as long day conditions with a lower light intensity if the total daily input of light energy is the same (Craker and Seibert 1982, Holmes and Smith 1977a).

A high light intensity will increase the oil concentration in numerous essential oil producing plants: peppermint (Hornok 1978, Clark and Menary 1980a), chamomile (Saleh 1973), American pennyroyal (*Hedeoma drummondii* L.) (Firmage 1981), thyme (*Thymus vulgaris* L.) (Yamaura *et al.* 1989), and dill (Hälvä *et al.* 1992a). On the other hand, Lincoln and Langenheim (1978) have observed in experiments using 'yerba buena' (*Satureja douglasii* (Benth.) Briq) that light levels scarcely influence oil concentration but will alter the oil composition. Firmage (1981) reported that the effect of light intensity depends on the genetic composition of plants. The effect of light intensity on the oil concentration may result from changes in enzyme activity, precursor formation or flow, and in accumulation and degradation of oil compounds (Bernáth 1986).

The explanations on the effect of light intensity on individual oil compounds is based on relatively few individual experiments only. Burbott and Loomis (1967) and Clark and Menary (1979) have shown that the high light intensity under long day conditions enhanced the formation of menthone and decreased the accumulation of menthofuran and pulegone in peppermint. Clark and Menary (1980a) reported, however, that menthofuran and also menthol contents are higher at low photon fluence rates. They also observed that high photon fluence rates favored limonene and cineole accumulation, especially under long day conditions. According to Saleh (1973), the amount of chamazulene in chamomile flowers will increase with increasing light intensity.

4.2.1. Effect of Light Quality

The light quality can modify the plant growth and development in numerous ways. The photochemical reactions in plants require energy from different parts of the light spectrum. Plants detect the light with chemical photoreceptors. Photomorphogenic pigments include phytochrome and blue light absorbing photoreceptors. The primary receptor that detects the quality of light is phytochrome which exists in two forms – Pr (absorption maximum at 660 nm) and Pfr (absorption maximum at 730 nm). Phytochrome action is characterized by red – far-red reversibility. The inactive form Pr will convert to the active form Pfr under exposure to red light (640–690 nm). The conversion from Pfr to Pr will occur under exposure to far-red light (710–750 nm). The phytochrome reversibility does work at high fluence rates (Holmes and Smith 1977b, Smith 1981, 1982). The mechanisms that might amplify the action of phytochrome include the control of gene function, the control of enzyme activity, the control of the hormone levels, and the control of membrane functions (Smith 1981, 1982).

Red and far-red light alone are known to cause more than a hundred responses in plants. The response to light quality is determined by the last irradiation before the dark period. The light intensity for the phytochrome action depends on the duration of irradiation and on the fluence rate. Phytochrome action is effective after a short exposure to red or far-red light. Certain responses, however, require longer or repeated exposures. Phytochrome regulates photomorphogenic responses such as dormancy, germination, flowering, pigment development, and leaf expansion (Holmes and Smith 1977b, Smith 1981, 1982).

The response of plants to blue light can also be partly controlled by the phytochrome. A large number of reactions to blue light show an action maxima at 370–380 nm and at 400–500 nm. The blue light receptor (a carotenoid, or preferably, a flavin-type molecule) is primarily responsible for detecting the quantity of light. Plant responses affected by the blue receptor include inhibition of stem growth, phototropism, chloroplast development, and synthesis of pigments, proteins, and enzymes. The chlorophylls and pigments, such as xanthophylls and carotenoids act as photoreceptors in photosynthesis (Gressel 1980, Thomas 1981).

The proportional amounts of red and far-red light, particularly, is important in determining stem elongation (Holmes and Smith 1977a, McLaren and Smith 1978, Morgan and Smith 1976, Smith 1981, Thomas 1981). The shoot elongation under far-red light is probably related to the plant hormones which are affected by the light quality (Holmes and Smith 1977b, Tucker 1976). The elongation of stems that is caused by far-red light is reduced by high light intensity (Meijer 1971, Warrington and Mitchell 1976). The response of plants to the light quality is related to changes in plant growth hormones under various light spectra. The relationship is, however, not clear. The effect of light quality may be modified by changing the amount of growth hormones, changing the activity of the substance, or by changing the sensitivity of plant tissue to the hormones (Degreef and Fredericq 1983, Holmes and Smith 1977a, Tucker 1976).

Temperature also moderates the effects of light quality on plants (Rajan *et al.* 1970). Far-red light has been reported to increase dry matter (Child and Smith 1987, Rajan *et al.* 1970, Tucker 1976, Vince-Prue 1977), reduce leaf area (Holmes and Smith 1977b, McLaren and Smith 1978, Rajan *et al.* 1970), and branching (Child and Smith 1987, Tucker 1976, Vince-Prue 1977) in numerous plants.

Dry weight, height, and leaf area of plants tend to increase even under short exposure and low intensity red light conditions (Vince 1964). Contrary, blue light is known to reduce the growth of plants (Cosgrove 1982, Eskins *et al.* 1989, Mortensen and Stroemme 1987). Increasing blue light will also result in shorter internodes (Casal and Smith 1989a,b, Warrington and Mitchell 1976) in the violet, blue, and red parts of the light spectrum. Plants were taller and produced a greater fresh biomass than those grown under lights which were richer in the green and yellow parts of the spectrum. The greatest number of shoots, however, developed under the light spectrum of 510–610 nm (green, yellow, orange). Saleh (1972) reported that chamomile produced the most flower heads and the greatest total of plant dry weight under white light

(34% at 400–510 nm, 50% at 510–610 nm). He also observed that red light increased the flower production more than blue and green light exposures. The size of the flower heads was, however, larger under green light.

The quality of end-of-day light did not affect the biomass production of dill. Regardless of the light quality treatments, all the plants received the same amount of photosynthetic light during the day (prior to the end-of-day treatments), and, presumably, the equal exposure to photosynthetic radiation resulted in the similar biomass production under the different light spectra (Hälvä et al. 1992). It seems, then, that the mechanism by which the light quality affected the oil concentration is other than the biomass production.

The effect of light quality on the oil production has been studied on chamomile (Saleh 1972, Verzar-Petri et al. 1978) and dill (Hälvä et al. 1992b). Saleh (1972) observed that white fluorescent light gave the highest total oil concentration and content of chamazulene in the flower heads of chamomile. More total oil and chamazulene per flower heads accumulated under green supplementary lights than under red or blue supplementary lights. However, the red light resulted in the highest total oil and chamazulene contents per plant due to the greater number of flowers and greater dry weight. Oil yield was smallest in the plants that received blue supplementary light. Conversely, Verzar-Petri et al. (1978) reported no qualitative differences in the essential oil of chamomile under various light spectra. The changes in the oil content reflect changes in the size of flower heads caused by the light quality.

In dill, far-red light, and to a smaller extent red light, increased the oil concentration. The effect of blue light was the opposite. Blue light affects the ontogenesy of plants. The plants grown under blue light flower later than the plants grown under red and far-red lights. Since all the plants were harvested at the same time, the plants grown under blue light were at an earlier developmental stage at the harvest than the plants grown under red and far-red light. The earlier developmental stage could account for the lower oil concentration in the plants grown under the blue light (Hälvä et al. 1992b).

4.2.2. The Effect of Daily Light Period

The net assimilation rate and the relative growth rate in several herb plants will increase in linear proportion to increasing light period. The length of the dark period can, however, be more important for plant responses than the length of the light period (Clark and Menary 1979, 1980a, Skrubis and Markakis 1976, Suchorska et al. 1988). Long-day plants are induced to ealier flowering under a long daily photoperiod. This photo-induction of flowering may result in lower vegetative herb yields under long day conditions (Franz et al. 1986, Putievsky 1983a,b).

Caraway grows taller and produces higher yields under short day conditions, while the daylength does not affect the yield of other fruit producing herb, coriander (Coriandrum sativum L.). Also, the size and number of caraway fruits per plant are greater under a short light period (Putievsky 1983b).

The early studies on dill showed that the fruit yield was the greatest under long day conditions, particularly when combined with low day/night temperatures (18/12°C).

Dill is known to flower and produce fruit even under continuous short day conditions although the plant is a long day plant (Hamner and Naylor 1939).

Basil (*Ocimum basilicum* L.) and oregano (*Origanum vulgaris* L.), both long day plants, produce higher herb yields under long day conditions (Skrubis and Markakis 1976, Putievsky 1983a). Skrubis and Markakis (1976) reported the highest basil yields with a 24 h light period, but high yields were also produced with 15–18 h light periods even under shorter growing times than under continuous light. According to Putievsky (1983a), daylength has no effect on the height of basil and oregano though most plants exhibit longer internodes and elongated stems when exposed to long days conditions.

Oil production increases with increasing light period in several herb plants: in peppermint (Burbott and Loomis 1967, Clark and Menary 1979 and 1980a, Grahle and Höltzel 1963), chamomile (Saleh 1968), tarragon (*Artemisia dracunculus* L.) (Suchorska *et al.* 1988), thyme (Yamaura *et al.* 1989), and dill (Hälvä *et al.* 1993).

Changing the photoperiod affects various compounds in different way. Under long day conditions, the amounts of menthol and menthone, major components of peppermint oil, will increase, while the amount of menthofuran, an undesirable product, will decrease (Burbott and Loomis 1967, Clark and Menary 1979 and 1980a, Franz *et al.* 1975, 1984, 1986, Grahle and Höltzel 1963). Accordingly, peppermint would produce more desirable oil composition when grown on higher latitudes compared to southern locations where the daily photoperiod is shorter than in the north. Under long day conditions low light intensity will produce oil, the contents of which are typical to short day conditions (Langston and Leopold 1954, Burbott and Loomis 1967). Furthermore, Clark and Menary (1979, 1980a) have reported a decrease in limonene, pulegone, and menthyl acetate, and an increase in cineole, b-pinene, sabinene, and trans-sabinene hydrate in peppermint oil under long day conditions.

Burbott and Loomis (1967) have suggested that the changes in oil composition reflect the balance between daytime photosynthesis and nighttime utilization of photosynthetic products, and they conclude that the photosynthesis, not the photoperiod, directly affects the oil production. Results of Clark and Menary (1980b) concerning the major constituents in peppermint oil support the direct effect of photosynthesis proposed by Burbott and Loomis. They, however, do not explain the changes in several minor constituents. Conversely, an identical oil composition in peppermint grown both under long day conditions and under simulated long day conditions (a short day with night interruption), implies a direct effect of photoperiod on the concentration and composition of essential oil (Clark and Menary 1979, Grahle and Höltzel 1963).

Oil production in chamomile tends to be rather stable under various environmental conditions. The terpene production, however, has a strong genetic control (Franz *et al.* 1986, Sváb *et al.* 1967). The chamazulene content in the oil is higher under long day conditions than under short day conditions (Franz *et al.* 1975, Saleh 1968). The proportions of bisaboloids in the oil also vary under different photoperiods (Franz *et al.* 1975, 1986). The composition of basil oil is clearly affected by photoperiod, though the effect on total oil production is unclear (Skrubis and Markakis 1976). The amount of linalool, the most abundant constituent in the basil oil, is high and rather constant under both short (9–12 h) and long (18–24 h) light periods but drops sharply when the light period exceeds 12 h.

In the environmental chambers, the oil concentration and biomass production of dill showed an inverse relationship in response to the photoperiod: a long light period increased the oil concentration while a short light period increased the herb yield. A high temperature seemed to compensate for the shorter light period, and the combination of a high temperature and a short light period produced the highest oil concentration. Conversely, under a long light period, a lower temperature was enough to produce the higher oil concentration in fresh herb (Hälvä et al. 1993).

The effect of photoperiod on essential oil production has also been studied by determining the number of trichomes in oil-producing plants. The total amount of monoterpenes correlated with the number of glandular trichomes in cotyledons and primary leaves in thyme (Yamaura et al. 1989). Also, the number of oil glands per unit area of lower epidermis in peppermint leaves (Langston and Leopold 1954) and in germander (Teucrium chamaedrys L.) (Bedaux 1952) increased under long day conditions increasing the total oil production.

4.3. EFFECT OF TEMPERATURE

The optimum temperature for the growth of each species varies widely depending on genotype and the original natural habitat of the species (Franz et al. 1984, Herath et al. 1979, Lincoln and Langenheim 1978, Putievsky 1983a). Accordingly, basil, native to warm climates, accumulates the greatest yields under warm conditions (30°C) while biennial caraway and oregano, naturally found in colder climates, produce the largest yields under moderate (18/24°C) temperatures (Putievsky 1983a). For the biennial caraway, low temperatures (16/20°C) at flowering and seed binding stages are most favorable. On the other hand, the annual caraway which is mainly grown in the Mediterranean region requires more heat for seed production. The plant does not commonly produce fruits when grown at the northern locations (Hälvä et al. 1986, Sváb 1992). However, a fruit yield was harvested in Canada but the content of the main compound, carvone was lower than desired by the industry. The low carvone content is attributed to the fruit that was not wholly ripe at the harvest due to the short growing season (Wahab 1997). In this case the irrigation clearly increased the fruit yield as also reported earlier in plants grown in Israel (Putievsky 1978). Bernáth and Hornok (1992) concluded in their review that the influence of temperature is relative. Accordingly, the content of active substances will either increase or decrease depending on the optimum temperature for the given species.

The optimum temperature will decrease at low light intensity. The optimum temperature for the leaf development and essential oil concentration in peppermint is approximately 21°C (Biggs and Leopold 1955, Clark and Menary 1980b). The optimum temperature for flower head and oil production in chamomile is 15°C. Also, a constant high temperature (25°C) favored the essential oil production in chamomile (Saleh 1970). In Japanese mint (Mentha arvensis L. var. piperascens Holmes), a combination of high day and low night temperatures (35/13°C) produced the greatest number of leaves and the highest amount of essential oil per plant. A day temperature of 30°C, however, produced the maximum dry weight regardless of the night temperature. At lower day temperatures, the oil production increases with increasing night temperature (Duriyaprapan *et al.* 1986).

In cold conditions, the growth rate is slow and the plants become more branched than under warm conditions. In pennyroyal, also more flowers will develop at each node at low temperatures (Firmage 1981). According to Saleh (1970), chamomile will produce more but smaller flower heads at high temperatures.

The difference in oil content in herbs is sometimes explained by the so called latitude effect. Temperature and day length cause the main climatic differences between the southern and northern locations. The carvone/limonene ratio in wild caraway in Finland is found to be higher at northern latitudes compared to southern latitudes. The same applies to the ratio on higher versus lower elevations in the Alps (Galambosi and Peura 1996). The term latitude-effect, however, seems to be too general to describe any variation in the growth and oil concentration of herbs as shown in the case of caraway and dill (Hälvä *et al.* 1986, Hälvä 1993). The plants were grown on similar subsrates in northern and southern locations. The growing site did not explain the variation in the oil content.

The factors that affected the fresh herb production the most, were the number of degree days and the amount of rainfall. Also, warm conditions increased the essential oil concentration and the contents of major oil components in dill. The latitude of the growing alone site did not explain the variation in the herb and oil production (Hälvä *et al.* 1988). In the growth chambers, contrary to the field conditions, dill yield tended to be highest under cool conditions. Under cool conditions, the growth period was longer, and the longer growth period could possibly have helped to accumulate more plant tissue than the shorter growth periods under warmer conditions. Under cool conditions, low respiration can preserve the assimilates and lead to greater biomass production. Also, the relatively low light intensity in the chambers may have affected the result, because the optimum temperature for plants is known to decrease at low light intensities.

The oil concentration in dill decreased with increasing fresh biomass production and tended to correlate with dry matter production and the mean daily temperature. The correlation between the dry matter production and oil concentration, again, implies a close relation between oil production and photosynthesis. The fresh herb yields decreased the further north the dill was grown. The low yields in the north were due to less favorable soil and cooler weather. Even though the fresh herb yield seemed much greater in the southern locations, the oil concentration was, in fact, unaffected by the latitude (Hälvä 1993).

Franz *et al.* (1984) have reported that temperature does not significantly affect the oil concentration in peppermint whereas Clark and Menary (1980b) have observed that a high day temperature tends to increase the oil accumulation. Oil concentrations both in peppermint and Japanese mint increase in linear proportion with increasing dry matter production and with increasing mean daily temperatures (Burbott and Loomis 1967, Duriyaprapan *et al.* 1986). A similar relation between oil production and dry weight has been reported in chamomile under various temperatures (Saleh 1970).

Though night temperature has little effect on total oil concentration in peppermint, it does significantly affect the oil composition. Also, an interaction between temperature and photoperiod affects the proportion of compounds. Short warm days combined with warm nights favor pulegone and menthofuran, while cool nights favor the formation of menthone. Under long days, the effect of night temperature on the oil composition in peppermint is less important (Burbott and Loomis 1967). According to Firmage (1981), american pennyroyal accumulates the largest concentration of menthol and isomenthone at cold conditions (9°C) even though the total oil concentration decreased at low temperatures.

The oil concentration in chamomile is highest at constant high temperature and decreases in linear proportion with decreasing temperature. The total oil production as well as chamazulene content per plant are, however, greatest under cool temperatures due to increased size of flower heads. The maximum chamazulene production will occur at constant moderate (20°C) temperature. In another experiment, cool nights (15°C) with day temperatures of 25°C produced both the highest total oil and chamazulene contents (Saleh 1970). The effect of temperature is controlled by the genotype of the plant as shown in the case of chamazule in chamomile oil (Sváb et al. 1967).

Burbott and Loomis (1967) have observed an increase in the synthesis of reduced compound at low temperatures and they suggested that the increase may be attributed to decreased respiration. Warm nights decrease the amounts of respiratory substrates (oxidizing conditions) while cool nights preserve higher amounts of respiratory substrates (reducing conditions). The oxidation – reduction level of monoterpenes reflects the oxidation – reduction state of the respiratory co-enzymes of the monoterpene-producing cells, which, in turn, depends on the concentration of respiratory substrates in the cells.

This theory has been later supported by Clark and Menary (1980b). They have reported that high light intensity (>500 μmol m-2 s-1), cool nights, and a day temperature of 20°C will maintain high levels of photosynthetic products which favor the reduction of pulegone to menthone in peppermint. Duriyaprahan et al. (1986), on the other hand, reported that the menthone content in Japanese mint will increase under high temperatures and that the menthol content tends to be stable under various temperatures.

The optimum temperature of 16–18°C is reported for the oil production of coriander, and 19–21°C for lavender (Lavandula angustifolia Mill.) (Savchuk 1976). Huopalahti (1984) has concluded that the oil concentration in dill herb will increase with decreasing temperature. The same applies to the content of anethofuran (i.e. the benzofuranoid), which is known to be responsible for the dill aroma. Lincoln and Langenheim (1978) reported that a low temperature (15°C) favors the oil production of 'yerba buena'. Low temperatures also increased the oil concentration in peppermint (Rabak 1916, Biggs and Leopold 1955), lavender (Savchuk 1976), and citronella (Cymbopogon nardus (L.) Rendle) (Herath et al. 1979).

The effect of temperature has been studied by counting the number of oil glands. Increasing temperatures clearly increased the number of glands in peppermint leaves (Biggs and Leopold 1955). The number of oil glands is, however, not necessarily proportional to the oil concentration because the oil will evaporate more rapidly at high temperatures (Hockings and Edwards 1943, Duriyaprahan et al. 1986). Accordingly, in Japanese mint, the number of oil glands was greater at high day temperatures regardless of the night temperatures, but there was no obvious correlation between the number of oil glands and oil production (Duriyaprahan *et al.* 1986).

4.4. SUMMARY

The species, subspecies, and the whole genetic composition of the plant will clearly affect how the plant will respond to the various ecological conditions. Furthermore, the developmental stage of the plant will influence the plant's response to the environment. Relatively few studies on various species show somewhat conflicting results. Light and temperature are intimately related in nature, and neither of them can be isolated from the other climatic factors. The studies have shown that the fruit and essential oil yield of caraway largely depends on the availability of the carbohydrate assimilates in early stages of seed production. The plant requires a high light intensity and a high rate of photosynthesis in order to produce good quality and great quantity of essential oil. Also a short photoperiod will contribute to greater size and number of fruits in caraway. The effect of light quality and the mechanism by which the light affects the production of caraway still remain to be discovered. The effect of photoperiod in nature is closely linked to the temperature. The optimum temperature for the species controls the level of oil production. The biennial caraway favors moderate temperatures while the annual caraway prefers higher temperatures for good growth. In general, the maximum amount of metabolites are synthesized under conditions that are most favorable for overall plant growth.

REFERENCES

Bedaux, F.C. (1952) Experimenteel onderzoek naar de variabiliteit van de stomataindex en een index voor Labiatenklieren van *Teucrium chamaedrys*. *Pharmaceutisch Weekbl.*, **87**, 652–657.

Bernáth, J. (1986) Production ecology of secondary plant products. In L.E. Craker and J.E. Simon (eds.) *Herbs, spices, and medicinal plants: Recent advances in botany, horticulture, and pharmacology* Vol. 1, Oryx Press, Phoenix, Arizona, pp. 185–234.

Bernáth, J. and Hornok, L. (1992) Environmental factors. In L. Hornok (ed.) *Cultivation and processing of medicinal plants*, Wiley and Sons. Chister, pp. 56–68.

Bernáth, J. and Tétényi, P. (1978) The effect of environmental factors – adaptibility relationship of stereoid alkaloid production based on investigation of two species, *Solanum laciniatum* Ait. and *Solanum dulcamara* L. *Acta. Bot. Acad. Sci. Hung.*, **24**(1–2), 41–55.

Biggs, R.H. and Leopold, A.C. (1955) The effects of temperature on peppermint. *Proc. Am. Soc. Hort. Sci.*, **66**, 315–321.

Bouwmeester, H. and Kuijpers, A.M. (1993) Relationship between assimilate supply and essential oil accumulation in annual and biennial caraway (*Carum carvi* L.). *J. Essent. Oil Res.*, **5**, 143–152.

Bouwmeester, H.J., Davies, J.A.R., Smid, H.G. and Welten, R.S.A. (1995a) Physiological limitations to carvone yield in caraway (*Carum carvi* L.). *Ind. Crops Products*, **4**(1), 39–51.

Bouwmeester, H.J., Smid, H.G. and Loman, E. (1995b) Seed yield in caraway (*Carum carvi* L.) 2. Role of assimilate availability. *J. Agric. Sci.*, **124**(2), 245–251.

Burbott, A.J. and Loomis, W.D. (1967) Effects of light and temperature on the monoterpenes of peppermint. *Plant Physiol.*, **42**, 20–28.

Casal, J.J. and Smith, H. (1989a) The end-of-day phytochrome control of internode elongation in mustard: kinetics, interaction with the previous fluence rate, and ecological implications. *Plant Cell Environ.*, **12**, 511–520. Casal, J.J. and Smith, H. (1989b) Effect of blue light pretreatment on internode extension growth in mustard seedlings after transition to darkness: analysis of the interaction with phytochrome. *J. Exp. Bot.*, **40**(217), 893–899.

Child, R. and Smith, H. (1987) Phytochrome action in light-grown mustard: kinetics, fluence-rate compensation and ecological significance. *Planta*, **172**, 219–229.

Clark, R.J. and Menary, R.C. (1979) Effects of photoperiod on the yield and composition of peppermint oil. *J. Amer. Soc. Hort. Sci.*, **104**(5), 699–702.

Clark, R.J. and Menary, R.C. (1980a) Environmental effects on peppermint (*Mentha piperita* L.). I. Effect of daylength, photon flux density, night temperature, and day temperature on yield and composition of peppermint oil. *Aust. J. Plant Physiol.*, **7**, 685–692.

Clark, R.J. and Menary, R.C. (1980b) Environmental effects on *peppermint* (*Mentha piperita* L.). II. Effect of temperature on photosynthesis, photorespiration, and dark respiration in peppermint with reference to oil composition. *Aust. J. Plant Physiol.*, **7**, 693–697.

Cosgrove, D.J. (1982) Rapid inhibition of hypocotyle growth by blue light in *Sinapis alba* L. *Plant Sci. Lett.*, **25**, 305–312.

Craker, L.E. and Seibert, M. (1982) Light energy requirements for controlled environment growth of lettuce and radish. *Transactions ASAE*, **25**(1), 214–216.

Croteau, R. (1986) Biochemistry of monoterpenes and sesquiterpenes of the essential oils. In L.E. Craker and J.E. Simon, (eds.) *Herbs, spices, and medicinal plants: Recent advances in botany, horticulture, and pharmacology* Vol. 1, Oryx Press. Phoenix, Aizona, pp. 81–133.

Degreef, J.A. and Fredericq, H. (1983) Photomorphogenesis and hormones. In W. Shoropshire and H. Mohr (eds). *Photomorphogenesis*, Encycl. Plant Physiol. NS16A. Springer Verlag, Berlin, Germany, pp. 401–427.

Duriyaprapan, S., Britten, E.J. and Basford, K.E. (1986) The effect of temperature on growth, oil yield and oil quality of Japanese mint. *Ann. Bot.*, **58**, 729–736.

Eskins, K., Westhoff, P. and Beremand, P.D. (1989) Light quality and irradiance level interaction in the control of expression of light-harvesting complex of photosystem II. *Plant Physiol.*, **91**, 163–169.

Evans, L.T. (1973) The effect of light on plant growth, development and yield. *Proc. Uppsala Symp. Plant response to climatic factors*, Ecology and conservation 5. UNESCO, Paris, pp. 21–35.

Firmage, D.H. (1981) Environmental influences on the monoterpene variation in Hedeoma drummondii. *Biochem. Sys. Ecol.*, **9**, 361–383.

Flück, H. (1955) The influence of climate on the active principles in medicinal plants. *J. Pharm. Pharmacology*, **7**, 361-383.

Franz, Ch., Ceylan, A., Hölzl, J. and Vömel, A. (1984) Influence of the growing site on the quality of *Mentha piperita* L. - oil. *Acta Hort.*, **144**, 145–150.

Franz, Ch., Fritz, D. and Schröder, F.J. (1975) Einfluss ökologischer Faktoren auf die Bildung des ätherischen Öls und der Flavone verschiedene Kamillenherkünfte 2. Einfluss von Licht und Temperatur. *Planta Med.*, **27**, 46–52.

Franz, Ch., Hårdh, K, Hälvä, S., Müller, E., Pelzmann, H. and Ceylan, A. (1986) Influence of ecological factors on yield and essential oil of chamomile (*Chamomilla recutita* (L.) Rauschert syn. *Matricaria chamomilla* L.). *Acta Hort.*, **188**, 157–162.

Galambosi, B. and Peura, B. (1996) Agrobotanical features and oil content of wild and cultivated forms of caraway (*Carum carvi* L.). *J. Ess. Oil Res.*, **8**, 389–397.

Grahle, A. and Höltzel, C. (1963) Photoperiodische Abhängigkeit der Bildung des ätherischen Öls bei *Mentha piperita* L. *Naturwissenschaften*, **50**, 552.

Gressel, J. (1980) Blue light and transcription. In H. Senger, (ed.), *The blue light syndrome*, Springer-Verlag, Berlin, pp. 133–153.

Hamner, K.C. and Naylor, A.W. (1939) Photoperiodic responses of dill, a very sensitive long day plant. *Bot. Gaz.*, **100**, 853-861.

Hälvä, S. (1993) Effect of light and temperature on the growth and essential oil of dill (*Anethum graveolens* L.). + *App. Academic Dissertation*, Cambridge, Massachusetts, p. 56.

Hälvä, S., Huopalahti, R., Franz, Ch. and Mäkinen, S. (1988) Herb yield and essential oil of dill (*Anethum graveolens* L.) at different locations. *J. Agric. Sci. Finl.*, **60**, 93–100.

Hälvä, S., Craker, L.E., Simon, J.E. and Charles, D.J. (1992a). Light levels, growth, and essential oil in dill (*Anethum graveolens* L.). *J. Herb, Spices, and Medicinal Plants*, **1**(1/2), 47–58.

Hälvä, S., Craker, L.E., Simon, J.E. and Charles, D.J. (1992b) Light quality, growth, and essential oil in dill (*Anethum graveolens* L.). *J. Herb, Spices, and Medicinal Plants*, **1**(1/2), 59–69.

Hälvä, S., Craker, L.E., Simon, J.E. and Charles, D.J. (1993) Growth and essential oil in dill (*Anethum graveolens* L.) in response to temperature and photoperiod. *J. Herb, Spices, and Medicinal Plants*, **1**(3), 31–39.

Hälvä, S., Hirvi, T., Mäkinen, S. and Honkanen, E. (1986) Yield and glycosinolates of mustard seeds and volatile oil of caraway and coriander fruit. II Yield and volatile oil of caraway fruit (*Carum carvi* L.). *J. Agric. Sci. Finl.*, **58**, 163–167.

Herath, H.M.W., Iruthyathas, E.E. and Ormrod, D.P. (1979) Temperature effects on essential oil composition of citronella selections. *Econ. Bot.*, **33**(4), 425–430.

Hockings, G.M. and Edwards, L.D. (1943) The utility of determination of numbers and dimensions of grandular scales in *Mentha species. J. Amer. Pharm. Assoc.*, **32**, 225–231.

Holmes, M.G. and Smith, H. (1977a) The function of phytochrome in the natural environment. I. Characterization of daylight for studies in photomorphogenesis and photoperiodism. *Photochem. Photobiol.*, **25**, 533–538.

Holmes, M.G. and Smith, H. (1977b) The function of phytochrome in the natural environment. IV. Light quality and plant development. *Phytochem. Photochem. Photobiol.*, **25**, 551–557.

Hornok, L. (1978) *Gyógynövények termesztése és feldolgozása* (Production and processing of medicinal plants). Mezőgazdasági Kiadó, Budapest.

Huopalahti, R. (1984) Effect of latitude on the composition and content of aroma compounds in dill, *Anethum graveolens* L. *Lebensm. - Wiss. Technol.*, **17**, 16–19.

Langston, R. and Leopold, A.D. (1954) Photoperiodic responses of peppermint. *J. Amer. Soc. Hort. Sci.*, **63**, 347–352.

Lincoln, D.E. and Langenheim, J.H. (1978) Effect of light and temperature on monoterpenoid yield and composition in *Satureja douglasii. Biochem. Sys. Ecol.*, **6**, 21–32.

McLaren, J.S. and Smith, H. (1978) Phytochrome control of the growth and development of Rumex obtusifolius under simulated canopy light environments. *Plant Cell Environ.*, **1**, 61–67.

Meijer, G. (1971) Some aspects of plant irradiation. *Acta Hort.*, **22**, 103–108.

Morgan, D.C. (1981) Shadelight quality effects on plant growth. In H. Smith (ed.) *Plants and daylight spectrum*, Academic Press, London, pp. 205–221.

Morgan, D.C. and Smith, H. (1976) Linear relationship between phytochrome photoequilibrium and growth in plants under simulated natural radiation. *Nature*, **262**, 210–211.

Morgan, D.C. and Smith, H. (1981) Non-photosynthetic responses to light quality. In J. Grace, E.D. Ford and P.G. Jarvis, (eds.) *Plants and their atmospheric environment*, Blackwell Scientific Publications, Oxford, pp. 109–134.

Mortensen, L.M. and Stroemme, E. (1987) Effect of light quality on some greenhouse crops. *Sci. Hortic.*, **33**, 27–36.

Putievsky, E. (1978) Yield components of annual *Carum carvi* L. grown in Israel. *Acta Hort.*, **73**, 283–287.

Putievsky, E. (1983a) Temperature and day-length influences on the growth and germination of sweet basil and oregano. *J. Hort. Sci.*, **58**(4), 583–587.

Putievsky, E. (1983b) Effect of day-length and temperature on growth and yield components of three seed spices. *J. Hort. Sci.*, **58**(2), 271–275.

Rabak, F. (1916) The effect of cultural and climatic conditions on the yield and quality of peppermint oil. *U.S. Dept. Agr. Prof.*, Paper 454.

Rajan, A.K., Betteridge, B. and Blackman, G.E. (1970) Interrelationships between the nature of the light source, ambient air temperature, and the vegetative growth of different species within growth cabinets. *Ann. Bot.*, **35**, 323–343.

Saleh, M. (1968) Effects of light upon quantity and quality of *Matricaria chamomilla* L. – oil. I. Preliminary study of daylength effects under controlled conditions. *Meded. Landb. hogeschool. Wageningen*, **68**(21), 1–14.

Saleh, M. (1970) The effect of the air temperature and the thermoperiod on the quantity and the quality of *Matricaria chamomilla* L. *Meded. Landb hogeschool. Wageningen*, **70**(15), 1–17.

Saleh, M. (1972) Effect of light upon quantity and quality of *Matricaria chamomilla* L. oil. *Pharmazie*, **27**(9), 608–611.

Saleh, M. (1973) Effect of light upon quantity and quality of *Matricaria chamomilla* oil. III. Preliminary study of light intensity effects under controlled conditions. *Planta Med.*, **24**, 337–340.

Savchuk, L.P. (1976) Influence of weather conditions on the amounts of essential oils in lavender and coriander. *Meteorol. Gidrol.*, **7**, 94–99.

Skrubis, B. and Markakis, P. (1976) The effect of photoperiodism on the growth and the essential oil of *Ocimum basilicum* (sweet basil). *Econ. Bot.*, **30**, 389–393.

Smid, H.G. and Bouwmeester, H.J. (1993) *Effect of light and pollination on the seed set and oil synthesis in caraway*. Trial results for 1992. Verslag – CABO-DLO 181. 33 p. + App.

Smith, H. (1981) Light quality as an ecological factor. In J. Grace, E.D. Ford and P.G. Jarvis, (eds.), *Plants and their atmospheric environment*. Blackwell Scientific Publications, Oxford, pp. 93–110.

Smith, H. (1982) Light quality, photoperception, and plant strategy. *Ann. Rev. Bot. Physiol.*, **33**, 481–518.

Suchorska, K., Jedraszko, B. and Olszewska-Kaczynska, I. (1988) Effect of temperature and day-length on the synthesis of biologically active substances of some essential oil plants. *Warsaw Agric. Univ. Ann. Rep.* 1987, **22–23**, Warsaw Agric. Univ. Press.

Sváb, J. (1992) Caraway (*Carum carvi* L.). In L. Hornok (ed.) *Cultivation and processing of medicinal plants*, Wiley & Sons. Chister, pp. 154–159.

Sváb, J., El-Din-Awaad, C. and Fahmy, T. (1967) The influence of highly different ecological effects on the volatile oil content and composition in the chamomile. *Herba Hung.*, **6**(2), 177–188.

Tétényi, P. (1986) Chemotaxonomic aspects of essential oils. In L.E. Craker and J.E. Simon (eds.) *Herbs, spices, and medicinal plants: Recent advances in botany, horticulture, and pharmacology* Vol. 1, Oryx Press, Phoenix, Arizona, pp. 11–32.

Thomas, B. (1981) Specific effects of blue light on plant growth and development. In H. Smith, (ed.) *Plants and the day light spectrum*, Academic Press, London, pp. 443–459.

Toxopeus, H. and Bouwmeester, H.J. (1992) Improvement of caraway essential oil and carvone production in The Netherlands. *Ind. Crops Products*, **1**, 295–301.

Toxopeus, H., Lubberts, J.H., Neervoort, W., Folkers, W. and Huisjes, G. (1995) Breeding research and in vitro propagation to impropve carvone production of caraway (*Carum carvi* L.). *Ind. Crops Products*, **4**(1), 33–38.

Toxopeus, H. and Lubberts, J.H. (1994) Effect of genotype and environment on carvone yield and yield components of winter-caraway in The Netherlands. *Ind. Crops Products*, **3**, 37–42.

Tucker, D.J. (1976) Effect of far-red light on the hormonal control of side shoot growth in the tomato. *Ann. Bot.*, **40**, 1033–1042.

Verzar-Petri, G., Marczal, G., Lemberkovics, E. and Rajki, E. (1978) Morphological and essential oil production phenomena in chamomile growing in phytotron. *Acta Hort.*, **73**, 273–282.

Vince, D. (1964) Photomorphogenesis in plant stems. *Biol. Rev.*, **39**, 506–536.

Vince-Prue, D. (1977) Photocontrol of stem elongation in light-grown plants of Fuchsia hybrida. *Planta*, **133**, 149–156.

Wahab, J. (1997) Saskatchewan herb and spice industry. What's new? *Prairie Med. Arom. Plants Conf. Brandon Manitoba.* pp. 11–18.

Warrington, I.J. and Mitchell, K.J. (1976) The influence of blue- and red-biased light spectra on the growth and development of plants. *Agric. Meteorol.*, **16**, 247–262.

Yamaura, T., Takana, S. and Tabata, M. (1989) Light-dependent formation of glandular trichomes and monoterpenes in thyme seedlings. *Phytochem.*, **28**(3), 741–744.

5. QUESTIONS OF THE GENERATIVE DEVELOPMENT IN CARAWAY

ÉVA NÉMETH

Department of Medicinal Plant Production, University of Horticulture and Food Industry, Villányi str. 29-43, H-1114 Budapest, Hungary

5.1. THE INTRASPECIFIC FORMS OF CARAWAY

The length of vegetation period, based on the type of flower initiation seems to be the major ground in distinguishing two ecotypes, the biennial and the annual types (*Carum carvi* f. *biennis* and f. *annua*). While the biennial type is indigenous through Europe and had been known here since centuries, first experiences on the annual type can be dated back only to the beginning of the seventies this century. The European populations may have an Egyptian origin, where this type is well known and produced regularly.

However, a big row of handbooks does not mention the different forms and speaks only about caraway generally, even today. In practice, production of annual populations is growing and spreading in more and more countries. In Hungary, during the sixties the existence of the annual form was practically unknown (Kerekes 1967), it occupied 3–5% of the growing area ten years later (Sváb 1978) and it gives already the majority of the production – above 90% – in the nineties. In the recent years, introduction of the annual form into further countries to the north is taking place, cultivar development is known e.g. from The Netherlands and Germany (Toxopeus and Bouwmeester 1993, Pank and Quilitsch 1996).

Although there is a huge number of morphological, technological and chemical characteristics, according which the biennial and annual forms can be distinguished – height, branching, seed size, shattering, essential oil content, etc. –, the basic feature is the difference in the vegetation length. It is in tight connection with the frost tolerance and the need for vernalization process in flower initiation. The flowering behaviour and response of the plant to the ecological factors are manifested through physiological processes but are regulated genetically.

5.2. FACTORS IN FLOWER INITIATION AND DEVELOPMENT

In spite of crops of greater economical importance or even related vegetable species, caraway have not yet been investigated in detail for exact flowering induction requirement.

Factors of flower induction are in any case complex. A certain developmental phase is regularly a prerequisite of flowering. It is in tight connection with outside factors, – ecological conditions –, such as temperature and light, their value and duration.

5.2.1. The Role of Plant Development

5.2.1.1. Rootstock Diameter

The essential developmental stage for flowering of caraway plants is mostly characterised by the diameter of their rootstock. According to Glusenko (1977), on the base
of the leaf mass and the diameter of the rootstock the production of the stand can
be estimated. At least six leaves and a rootstock diameter of 5 mm are necessary at
the end of the first vegetation year for flowering in the next spring. Furthermore, he
proved a tight correlation between the mentioned two features (r: 0.92–0.99), so the
leaf number itself might give proper information about the plant development.
Observations of Havalda (1980) showed, that seed yield of a plantation depends on
the presence of plant individuals of at least 6 mm rootstock diameter. Between their
proportion in the stand and the mass of production a positive linear regression has
been described. Individuals of rootstock diameter more than 10 mm are of special
significance.

Similar results were published from Poland, by investigations of Weglarz (1982). He
described the advantage of the more developed individuals in flowering and yielding.
As a lower limit of flowering potential, he indicated an individual root mass of 26–50 g,
and a rootstock diameter of 8–15 mm. While among plants under this limit the flowering proportion was only 55–75%, in case of plants of the mentioned size it was
80–95%, and only the largest plants (50–150 g and 16–25 mm diameter) developed all
flowers (in 100%).

In case of the related *Daucus carota* L. even the effectiveness of artificial flowering
induction by GA3 was connected with plant size: while in 12 weeks old plants a maximum value of 84% flowering could be achieved, in 8 weeks old individuals 70% proportion was the best value (Bandara *et al.* 1995). *Apium graveolens* L. var. *dulce* may end
juvenility and become competent for effective cold treatment by getting 17–20 leaves,
depending on also the cultivar (genotype) (Ramin and Atherton 1991).

Plant size plays an important role both in making the individual capable of flowering
as well as developing more and larger flowers. In our experiment, carried out
in 1994–95 at the University of Horticulture and Food Industry, Department of
Medicinal Plant Production, caraway plants of different rootstock size were planted and
examined in phytotron chambers. Four different biennial origins were tested, in all
the four population the more developed plants had 9–13 mm rootstock diameter and
8–12 leaves, while the less developed ones had 5–8 mm rootstock diameter and
5–7 leaves. They were grown under two types of ecological programs. One of them
consisted of a two week, the other a seven week vernalization period (8°C at day and
5°C at night), after which the temperature raised from 14° to 21°C. Data of the experiment showed, that the larger plants did not require a long vernalization period, while in
case of the less developed roots only a longer induction may result in almost full flowering (Table 1). In general, – independent from genotype –, the thinner rooted plants could
never reach the flowering percentage of the thicker rooted ones (Németh *et al.* 1997).

In the average of origins and temperature treatments, the thick rooted plants exhibited also a higher and stronger growth. Especially the number of the stems and
that of the umbels were significantly higher in case of the more developed individuals

Table 1 Ratio of flowering individuals (in per cent) according to development and genotype after two weeks vernalization at 8/5°C

Taxon	2 weeks induction (Program 2)		7 weeks induction (Program 1)	
	Diameter of rootstock at the beginning of vernalization			
	5–8 mm	9–13 mm	5–8 mm	9–13 mm
Swiss	62	91	100	100
Finnish	71	100	90	100
Hungarian	75	100	92	100
Rumanian	88	100	98	100
Average	74.0	97.8	95	100

Table 2 Morphological characteristics dependent on rootstock size

Group	Plant height (cm)	Number of stems (Pcs)	Number of umbels (Pcs)	Diameter of main umbel (cm)	Diameter of secondary umbel (cm)	Diameter of stem (mm)
Thick rooted	48.5	1.8	20.1	6.0	4.6	5.0
Thin rooted	39.1	1.2	15.0	5.4	4.5	4.4
LSD	—	0.5	4.4	0.6	—	0.5
p%	—	5	5	10	—	10

(Table 2). The size of the umbels as well as the stem diameter also exceeded the values of the thin rooted plants (unpublished results).

5.2.1.2. Vegetation Length

Reaching the appropriate developmental stage is influenced by a lot of factors. The best known ecological and agrotechnical circumstances are: sowing time and rate, spacing, nutrient and water supply.

The role of plant size in flowering was proven by us in a *sowing time* optimalization experiment in Budapest, Soroksár. A special genotype – result of hybridisation between annual and biennial types – was sown into open field plots at two different time, at the beginning of August and at the beginning of September. Measurements at the end of October, before first frosts appearance, showed that plant development on the two plots was unequal. While plants which had been sown earlier, had a rootstock of 3–4 mm in diameter, the later sown population had only 1.5–2 mm. The number of leaves was also different, 6–7 leaves in the plots sown in August, and 4–5 leaves in the plots sown in September. Even in spring, in April, the earlier sown population exhibited significantly thicker roots (6–7 mm), than the other one (3–4 mm). As the main result of this growth difference, flowering percentage in the next year was also definitely different. Plant stand sown in August flowered in May fully, it means percentage of vegetative individuals remained under 5% (Figure 1). In the contrary, plant stand which was sown in September, exhibited a very sparse flowering with only some single

Figure 1 Experimental plots of overwintering caraway population (Budapest, 1997) left: plot sown in August, right: plot sown in September

stems (3–6%), (unpublished results). Obviously, the plants need a certain vegetation period to be able to reach a proper developmental stadium for flower production.

In Poland, traditional biennial cultivars were able to develop properly, when sowing was carried out till the end of May. In this case the plants could catch up with the stand sown in March. However, later sowings than the mentioned could not develop flowers in the next spring and no yield was obtained. Sowing time influenced the number of flowers, diameter of rootstock, fresh mass (Weglarz 1983a). Hungarian practice does not allow a sowing of biennial types later, than beginning of April (Sváb 1993). However, the reason of this strict rule may lay also in the poor precipitation supply later in the vegetation period, not only the length of the growing season itself. As an other example, Toxopeus (Chapter 7) mentions the end of June as the last possible sowing time in the more humid Dutch climate.

Further practical observations on the effect of *spacing* were made in 1994–96 at our experimental station. Gene-bank materials of biennial caraway had been studied for description and production evaluation. In the second examination circle accidentally a double seed rate was sown than usual. In this population we did not get any flowers, and of course no seed yield in the second vegetation year. It seems, that the small spacing prevented the plants in reaching the desired growth and flowering. The role of spacing in the generative development of caraway was mentioned already in earlier works (Kerekes 1967) and this feature studied and proved also in case of other essential oil crops (Németh and Bernáth 1992).

5.2.1.3. Nutrition and Water Supply

These are indirect factors in flowering of a stand. They are of basic importance for the plants to reach the appropriate size when flower initiation may take place. Weglarz (1983b) proved, that under the circumstances of this experiment, the highest rate of nitrogen, phosphorous and potassium fertilisers resulted in 60% flowering of the plant stand, while the control plot – without fertilisation – did not flowered at all in the second vegetation year. Differences in soil water capacity caused similar changes. Among the applied treatments 50–70% water capacity was shown as optimal.

5.2.2. The Role of Ecological Factors

5.2.2.1. Temperature

In a row of plant species, cold effect is the major factor stimulating flower initiation. Its value and length are of basic importance, however, they are satisfactorily cleared up only for a few species. As for caraway, literature sources has a common accepted, simply opinion: the biennial type does require a cold period for vernalization.

As for the value of effective temperatures, no exact data are known. Sváb (1993), similarly to other authors, only mentions a need for winter cold temperatures.

Also for the length of the effective vernalization period in caraway, the literature mentions only practical observations. According to Bouwmeester and Kuijpers (1993), biennial varieties can meet their vernalization requirement by the end of January. While it should be true under average weather conditions in The Netherlands, the whole winter is not enough for effective flower initiation of biennial varieties e.g. in Israel (Putievsky et al. 1994), where winter temperatures are moderate. Temperature values and their duration should be in a tight connection.

Furthermore, the required vernalization length is in connection with the development and size of the plants, more developed ones demanding a shorter period. Also other factors are involved, in several cases the interaction of temperature and photoperiods are proved or supposed (Ramin and Atherton 1994).

For the majority of the most important vegetables of the Umbelliferae family temperatures between 5–10°C proved to be the most effective, however both lower (5°C) and higher (15°C) temperatures might have an inductional effect (Rünger 1977). Optimal length of the inductive period however, seems to be different for each species. While carrot needs at least two weeks and parsnip 40–50 days of vernalization, celery gets vernalized even in some days, according to the mentioned author. However, Ramin and Atherton (1994) state, that optimal induction for *Apium graveolens* L. var. *dulce* should continue at least 21 days at a temperature of 5°C which is accompanied by short day photoperiod. Also differences in effective vernalization period among varieties of *Daucus carota* were shown by Dias-Tagliacozzo and Valio (1994), where cultivar 'Nantes' required at least 3 months of inductive temperatures, for 'Brasilia' 5 days proved to be sufficient. According to practical observations of Toxopeus (Chapter 7) biennial caraway varieties need about 8 weeks vernalization period at temperatures below 10°C.

It seems, that vernalization circumstances are characteristic not only for species but also for intraspecific taxa. The lack of information about these features especially in case of *Carum carvi* L. has a significant drawback in genetical and agricultural development and emphasises the importance of carrying on the necessary studies.

A row of experiments was carried out recently at our department in order to answer these questions. The idea began with a breeding experiment, carried out at our department in 1993, when biennial caraway individuals were transplanted from open field into containers in a phytotron chamber. The transplantation took place at the beginning of October. After two weeks growing period at 8°C, several plants started to develop flowering stems. Although it was not the original question of that experiment, the problem raised: might all those plants have got the suitable vernalization effect? Were the September temperatures in the field or even the 1–2 weeks phytotron programme enough for the flowering induction? Might it happen, that biennial varieties do not require a definite cold period at all, winter only stops their development for a while?

In order to be able to answer these question to some extent, a row of experiments were installed in 1994–96 in the frame of our university research work (Németh *et al.* 1997).

In phytotron chambers, 6 months old individuals of the biennial variety *Maud* had been grown for 11–16 weeks. Five temperature programs had been applied as treatments for flowering induction (Table 3). Differences of the programmes were installed mainly regarding of the length of the cold period. As "cold" a 8/5°C (day/night) programme was applied, which lasted one, two and seven weeks, respectively. As control programs, day/night temperatures of 15/10°C and 22/15°C were installed.

Table 3   Experimental programs for evaluation of temperature effect on caraway
flowering (day/night temperatures in °C)

Week	program 1	program 2	program 3	program 4	program 5
1	18/12	18/12	15/10	15/10	22/15
2	22/16	22/16	15/10	15/10	22/15
3	22/16	22/16	8/5	15/10	22/15
4	8/5	8/5	15/10	15/10	22/15
5	8/5	8/5	15/10	15/10	22/15
6	8/5	15/10	15/10	15/10	22/15
7	8/5	15/10	15/10	15/10	22/15
8	8/5	15/10	15/10	15/10	22/15
9	8/5	15/10	15/10	15/10	22/15
10	8/5	15/10	15/10	15/10	22/15
11	14/10	17/12	15/10	15/10	22/15
12	14/10	17/12			22/15
13	16/12				
14	16/12				
15	21/16				
16	21/16				

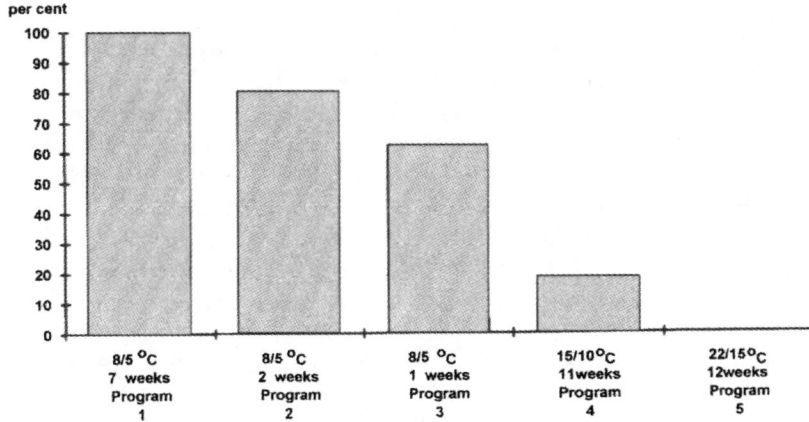

Figure 2 Proportion of flowering individuals after different vernalization treatments

It could be established, that biennial caraway really require some cold effect for generative differentiation. The results show, that the constant 22/15°C could not induce any flower development within the studied period. Lower temperature, as constant 15/10°C had a moderate induction effect, it stimulated flowering in nearly 20% of the individuals. The vernalization at 8/5°C day/night temperatures resulted in the highest proportions of flowering individuals, the longer this period, the more flowering plants (Figure 2). After the 7 week induction period the whole population developed stems. Thus, it seems, that by this latter treatment the optimal vernalization circumstances could be approached. According to the data, caraway optimal induction regime might lay between 5°C and 8°C, which is effective when lasting more than 2 weeks. Both a shorter period as well as higher temperatures up to 15°C result in partial flowering. However, lower effective temperatures can not be excluded on the basis of the existing results.

5.2.2.2. Illumination

Beside temperature, illumination length may also play a basic role in flower initiation (Rünger 1977). In case of several species those inductive factors stay in tight correlation with each other (Booij and Meurs 1994). The most frequent reaction types are: short day treatment, short day treatment before or during the cold period, long day treatment after a cold period, etc.

Illumination may act on flowering through its length during the day (number of light and dark hours), its length during the plant life (number of cycles of photoperiodic induction) and sometimes its intensity.

In case of caraway scientific data on photoperiodic reaction are very few. Putievsky (1983) examined the effect of daylength and temperatures on the flowering of three Umbelliferous species: caraway, dill and coriander. The three spices exhibited different reactions to the treatments. Caraway developed flowers under all the experimental circumstances (18/12°C or 24/12°C day and night temperatures, with 10 h or 16 h photoperiods). However, except plant height, the yield components (number of branches, seed weight, etc.) were all reduced by the long day treatment. Although flowering occurred about 3 weeks earlier on long day and higher temperatures, seed yield decreased by more than 30%. The experiment indicated, that caraway does not need any special photoperiod for flowering, however a longer vegetative growth at lower temperatures and short day may enhance generative mass.

As for the related Umbelliferous species, a reduction of time needed for flowering and an advantageous effect of long days on dry weight of the plant was established for the related dill (Hälvä *et al.* 1993). In *Apium graveolens* L. var. *rapaceum* short day treatment did not prevented flowering but did so for stem elongation (Booij and Meurs 1995). In case of *Foeniculum vulgare*, it was proved, that the critical photoperiod for umbel initiation is longer, than 13 hours. For 100% initiation at least 25 photoperiodic cycles are required (Peterson *et al.* 1993).

In our investigations biennial caraway individuals of different age were exhibited to a short day treatment. 10 hours dark periods and 14 hours light periods alternated during 28 days (Németh *et al.* 1997). The temperature during the whole treatment lie at 22/15°C (day/night). Four age groups were studied, which could be characterised by different sizes of the individuals. The smallest ones were 3 months old with 5.4 mm rootstock diameter, the others 4, 5, 6 months old with 7.0, 8.4 and 9.6 mm diameters respectively.

However, in neither plant group not any flowering stems appeared during the five months observation period. It might mean, either, that caraway does not need any short day induction for flower initiation at all, or any photoperiodic response is effective only with interaction of low temperatures. Besides, the success of a longer application of the mentioned treatment is not excluded.

5.3. GENETIC DETERMINATION OF INDUCTION REQUIREMENT

Although factors, which may induce flowering in caraway, are complex ones, genetical determination should be the basis of all the reactions. Differences in the vegetation length and induction requirement of biennial and annual types of caraway have their genetical background, which eventually should be the real difference between the ecotypes.

The genetical code regulates the physiological processes, through which the flower development takes place. According to Kuckuck *et al.* (1988) cross pollination – which is characteristic also for caraway –, is the basis for a heterogenous genepool of the species, and offer a possibility for recombination and natural selection. By this way a flexible adaptation potential to different ecological circumstances, first of all photoperiod and temperature is assured.

Until now little is known about inheritance of this mechanism, or even on the flower initialisation process, which is a problem with many question marks, and not only in case of caraway.

According to Putievsky *et al.* (1994) the lack of vernalization requirement – annual habit – may occur as a rare homozygote mutation of biennial caraway and can be stabilised by selection. Observations of the mentioned authors show, that selected individuals flowered already in the first year of growth, after which they remained alive and flowered anew during 2–3 vegetation periods. It is not the same as known for traditional annual varieties, because they die after seed ripening in the year of sowing.

Toxopeus and Lubberts (Chapter 7) discusses the simplest variation of the inheritance of vegetation length, which would explain the results of their own experiments. It is a single locus determination model with incomplete dominance of the allele for annual flowering. This model is indicated by segregation numbers in the F1 generations of hybrid strains between annual and biennial material, as well as that of the control populations.

In our hybridisation trials in Hungary, whose aim is the development of an annual, high quality cultivar, successive generations had been investigated to study the regulation of flowering in caraway and to try to define the main difference between the two ecotypes. First results on F1 and F2 had been published (Németh et al. 1996). Evaluation of further generations till now did not disprove the assumptions.

In this study, in F1 generation of annual and biennial crosses, 86–94% of the plant individuals flowered. There were no significant alterations between the reciprocal crosses, which is similar to findings of Toxopeus and co-workers (see above). It did not show a significant difference from control annual variety, where 100% flowering is also never reached.

However, the time of flower development during the vegetation period proved to be a characteristic feature. While accessions of mother plants in the annual cultivar *SZK-1* had a rather uniform blossoming, which lasted 3–4 weeks in June and July, progenies of mother plants of biennial cultivar *Maud* flowered from June till October.

In order of a proper evaluation the flowering proportion of the F2 generations, at the end of the vegetation period the vegetative individuals were also examined for flower formation by cutting the rootstocks. It was found, that further 20–65% individuals could be counted to the flowering group, because of the buds were observable although yet hidden in roots. In this manner, final flowering percentages showed a 3:1 proportion between flowering and vegetative plants. Similarly to the F1 populations, flowering was dragging on till the frosts, especially in case of accessions of biennial mother plants.

A possible hypothesis is the presence of several genes, among which a major gene effect may be observed. Presumably, the standard varieties *SZK-1* and *Maud* should be homozygote forms for the two alleles. According to this hypothesis, by crossing these genotypes, F1 plants are heterozygotes concerning this major gene (A), and thus, all of them are flowering in the first vegetation cycle, as found in our trial. In F2 the major gene segregates to 3:1 causing 25% of the population remaining vegetative in the year of sowing. In later generations, only collecting the seeds individually may result in stabilisation of annual habit, because of the big proportion of heterozygous plants.

Beside this regulation pattern, the presence of modifier gene/s (B) are most likely. The significant time shift in flowering of both the F1 as well as the F2 hybrid generations can be explained by this model. The distribution of flowering time allows different conclusions. One possible hypothesis about the effect of modifier genes is presented in Figure 3. In this case a slight difference in flowering time also between

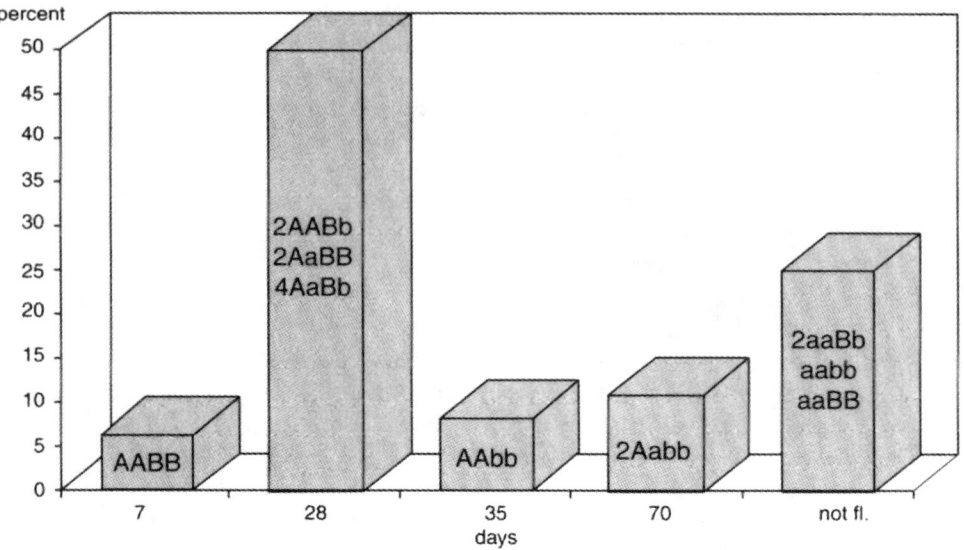

Figure 3 Distribution of F2 genotypes according to flowering time A-supposed major genes, B-supposed modifier genes (see text)

AA and Aa individuals is supposed. The modifier gene may act also as a gene with more alleles or, more likely have a polygenic character.

On the basis of our results, two physiological mechanisms for regulation of the vernalization process seems to be possible (Figure 4).

(a) Flowering inhibitors are present in both ecotypes, thus, dominant alleles influence the degradation process of them in annual caraway, while cold (vernalization) may have the same effect in biennial types. Flowering potential without vernalization should be a dominantly inherited feature e.g. in case of cereals (Bálint 1966).

(b) Inhibitors are only present in the biennial form, so the dominant allele acts here in breaking down them, being activated by cold treatment.

Cultivar differences in flowering time of different Umbelliferae species (Bhandari *et al.* 1991, Bernáth *et al.* 1996) might be caused by different amount of inhibitors. A similar system is present in sugar beet, where annual or perennial habit is regulated by a simple monogene (Kuckuck *et al.* 1988).

Thus, major genes in both cases might influence the presence of breaking down enzymes, and fit to our practical results. "Modifier" genes may influence the quantity of the inhibitors present or even the amount of hormones needed for flower initiation, so the speed of the generative process. Inheritance of this last feature shows a definite maternal effect.

However, the whole mechanism can only be imagined in tight correlation with other ecological and agrotechnical circumstances influencing plant development.

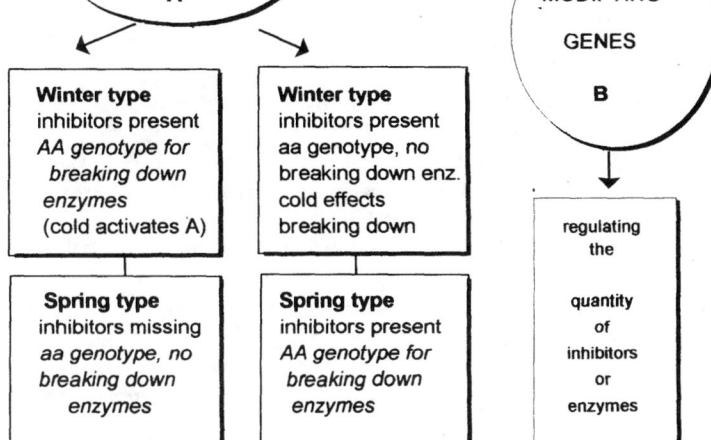

Figure 4 Models for the regulation of flowering induction in *Carum carvi* L.

REFERENCES

Bálint, A. (1966) *Breeding of agricultural crops* (*Mezőgazdasági növények nemesítése*), Mezödazdasági Publisher, Budapest. p. 193.

Bandara, P.M.S., Tanino, K.K., Nito, N., Looney, N.E., Nevins, D.J. and Halevy, A.H. (1995) Effect of gibberellic acid (GA3) on seedstalk development and flowering in carrot (*Daucus carota* L. var. *sativa* DC.). *Acta Horticult.*, **394**, 251–261.

Bernáth, J., Németh, É., Mihalik, E. and Kattaa, A. (1996) Stability and ecological flexibility of essential oil in *Foeniculum* taxa (Eltérő származású *Foeniculum* taxonok illóolajtartalmának stabilitása és ökológiai változékonysága). *Summaries of Scientific Symposium "Lippay"*, Budapest, October 1996. p. 136.

Bhandari, M.M., Adarsh, G. and Gupta, A. (1991) Variation and association analysis in coriander. *Euphytica*, **58**, 1–4.

Booij, R. and Meurs, E.J.J. (1994) Flowering in celeriac (*Apium graveolens* L. var. *rapaceum* (Mill.) DC): effects of photoperiod. *Scientia Horticulturae*, **58**, 271–282.

Booij, R. and Meurs, E.J.J. (1995) Effect of photoperiod on flower stalk elongation in celeriac (*Apium graveolens* L. var. *rapaceum* (Mill.) DC). *Scientia Horticulturae*, **63**, 143–154.

Bouwmeester, H.J. and Kuijpers, A. (1993) Relationship between assimilate supply and essential oil accumulation in annual and biennial caraway (*Carum carvi* L.). *J. Essent. Oil Res.*, **5**, 143–152.

Dias-Tagliacozzo, G.M. and Valio, I.F.M. (1994) Effect of vernalization on flowering of *Daucus carota* (cvs. Nantes and Brasilia). *Revista Brasileira de Fisiologia Vegetal.*, **6**, 71–73.

Glusenko, N.N. (1977) Results of the sudy on caraway collection (Resultati izucenija kollekcii tmina). *Annals of the Research Institute for Essential oil Crops of the Ministry of Agriculture* (*Trudy VNIIEMK*), Simferopol, USSR, pp. 57–62.

Halva, S., Craker, L.E., Simon, J.E. and Charles, D.J. (1993) Growth and essential oil in dill, *Anethum graveolens* L., in response to temperature and photoperiod. *J. of Herbs, Spices and Medicinal Plants*, 1, 47–56.

Havalda, Gy. (1980) Comparison experiments on annual and biennial taxa of caraway (Egy- és kétéves konyhakömény taxonok összehasonlító vizsgálata). *Diploma work at the University of Horticulture and Food Industry*, p. 48.

Kerekes, J. (1969) *Cultivation of medicinal plants* (*Gyógynövénytermesztés*), Mezőgazdasági Publisher, Budapest, pp. 147–151.

Kuckuck, H., Kobabe, G. and Wenzel, G. (1988) *Bases of plant breeding* (*a növénynemesítés alapjai*), Mezőgazdasági Publisher, Budapest, pp. 156–160.

Németh, É. and Bernáth, J. (1992) Investigations on flowering induction of *Salvia sclarea* populations (*Salvia sclarea* populációk virágzásindukció vizsgálata). *Special Issue of Gyógyszerészet as Proceedings of the 7th Hungarian Medicinal Plant Conference*, Székesfehérvár, p. 44.

Németh, É. and Pluhár, Zs. (1996) Preliminary observations on the inheritance of caraway flowering. *Beiträge zur Züchtungsforschung*, 2, 116–119.

Németh, É., Bernáth, J. and Pluhár, Zs. (1997) Factors influencing flower initiation in caraway (*Carum carvi* L.). *J. of Herbs, Spices and Medicinal Plants*, 5, 41–50.

Pank, F. and Quilitzsch, R. (1996) Phänotypische Variabilität des einjährigen Kümmels (*Carum carvi* L. var. *annuum* hort.) in mitteldeutschen Anbaugebiet. *Z. Arznei- und Gewürzpflanzen*, 1, 128–133.

Peterson, L.E., Clark, R.J. and Menary, R.C. (1993) Umbel initiation and stem elongation in fennel (*Foeniculum vulgare*) initiated by photoperiod. *J. Essent. Oil Res.*, 5, 37–43.

Putievsky, E. (1983) Effects of daylength and temperature on growth and yield components of three seed spices. *J. of Horticult. Science*, 58, 271–275.

Putievsky, E., Ravid, U., Dudai, N. and Katzir, I. (1994) A new cultivar of caraway (*Carum carvi* L.) and its essential oil. *J. Herbs, Spices and Medicinal Plants*, 2, 81–84.

Rünger, W. (1977) *Flower formation and development* (*Virágképződés és virágfejlődés*), Mezògazdasági Publisher, Budapest.

Ramin, A.A. and Atherton, J.G. (1991) Manipulation of bolting and flowering in celery (*Apium graveolens* L. var. *dulce*). II. Juvenility. *J. of Horticult. Science*, 66, 709–717.

Ramin, A.A. and Atherton, J.G. (1994) Manipulation of bolting and flowering in celery (*Apium graveolens* L. var. *dulce*). III. Effect of photoperiod and irradiance. *J. of Horticult. Science*, 69, 861–868.

Sváb, J. (1978) Caraway (Konyhakömény). In L. Hornok (ed.), *Cultivation and processing of medicinal plants* (*Gyógynövények termesztése és feldolgozása*), Mezògazdasági Publisher, Budapest, pp. 142–147.

Toxopeus, I.H. and Bouwmeester, J. (1993) Improvement of caraway essential oil and carvone production in The Netherlands, *Industrial Crops and Products*, 1, 295–301.

Sváb, J. (1993) *Carum carvi* L. In J. Bernáth (ed.), *Wild growing and cultivated medicinal plants* (*Vadon termô és termesztett gyógynövények*), Mezògazda Publisher, Budapest, pp. 179–183.

Weglarz, Z. (1982) Effect of agricultural agents on transition of *Carum carvi* L. from vegetative to generative phase I. Effect of seedling rootstock size on the value of caraway seedlings (Wplyw czynników agrotechnicznych na prechodzenie kminku zwyczajnego z fazy wegetatiwnej w generatiwna I. Wplyw wielkosci wysadkow na wartosc nasienników). *Herba Polonica*, 28, 171–177.

Weglarz, Z. (1983a) Effect of agricultural agents on transition of *Carum carvi* L. from vegetative to generative phase III. Effect of time of sowing, amount of seeds per ha and fertilization level on the development and cropping of caraway (Wplyw czynników agrotechnicznych na prechodzenie kminku zwyczajnego z fazy wegetatiwnej w generatiwna III. Wpływ terminu siewu, ilosci wysiewu nasion i poziomu nawozenia na rozwoj i plonowanie). *Herba Polonica*, **29**, 103–111.

Weglarz, Z. (1983b) Effect of agricultural agents on transition of *Carum carvi* L. from vegetative to generative phase II. Effect of fertilization and soil moisture on the development and cropping of *Carum carvi* L. (Wplyw czynników agrotechnicznych na prechodzenie kminku zwyczajnego z fazy wegetatiwnej w generatiwna II. Wplyw nawozenia i wilgotnosci gleby na rozwoj i plonowanie) *Herba Polonica.* **29.** 21–26.

6. REGULATION OF ESSENTIAL OIL FORMATION IN CARAWAY

HARRO J. BOUWMEESTER

Institute for Agrobiology and Soil Fertility (AB-DLO), P.O. Box 14, 6700 AA Wageningen, The Netherlands

6.1. INTRODUCTION

Caraway has been grown as a spice plant for many centuries and is almost cosmopolitan. It is native to Europe, parts of Asia and northern Africa, but it is grown also in the northern US and Canada. Caraway seeds technically are half-fruits, the whole fruit being a schizocarp which comprises two distinct halves ('mericarps') which each contain one seed. We will use 'seed' where we refer to the agricultural product (half fruit) and 'fruit' when we refer to the entire schizocarp (containing two seeds).

Caraway essential oil has been used as a flavouring for liquors and toothpaste, while the seeds have been used as a spice and flavouring. Only little attention has been paid to the improvement of caraway essential oil content and composition. The fact that caraway has only been a minor crop in most countries, has also contributed to this lack of attention. However, as part of a search for new industrial crops, in the early 90's in The Netherlands a research program on caraway was initiated. Several research projects were aimed at finding new, non-food industrial applications for caraway essential oil, in others the factors determining caraway essential oil yield and the possibilities to improve this yield were studied (breeding, disease control). The search for new uses focused on the biological activities that had been reported in literature for caraway essential oil or its main constituents, limonene and carvone (Figure 1). These biological activities include insecticidal or insect repellent effects (Hartzell 1944, Lichtenstein et al. 1974, Zuelsdorf and Burkholder 1978, Su 1985, 1987), antibacterial and antifungal effects (Koedam 1982b, Guérin and Réveillère 1985, Janssen et al. 1988), inhibition of seed germination (Asplund 1968) and inhibition of sprouting of potatoes (Beveridge et al. 1981).

The use of carvone as a natural sprouting inhibitor for potatoes proved to be the most interesting new application for caraway essential oil or rather one of its main constituents, carvone (Oosterhaven 1995, Hartmans, Chapter 13). The relative ease with which allowance was obtained for the use of (+)-carvone from caraway as a sprouting inhibitor in The Netherlands and probably soon in the entire EC is caused by the availability of extensive toxicological data already collected for the use of caraway as a spice. As a consequence, however, the carvone used to formulate the sprouting inhibitor has to meet specific requirements, such as high purity and enantiomeric excess. Particularly the latter demand makes a chemical synthesis of carvone (e.g. from limonene) difficult to be competitive to agricultural production with caraway which produces (+)-carvone

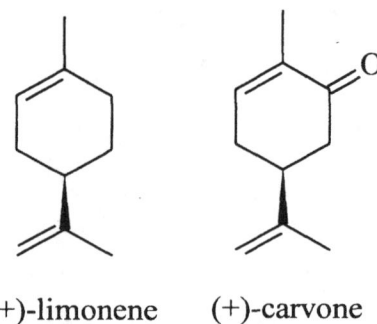

(+)-limonene   (+)-carvone

Figure 1   The major compounds of caraway essential oil

with an enantiomeric excess of over 99% (Bouwmeester *et al.* 1995b). However, the lack of research and breeding activities in the past now hamper large-scale introduction of carvone as sprouting inhibitor. Because of the low carvone yields of caraway the price of carvone is too high. Nevertheless, agricultural production of carvone, that is from a caraway (or dill) crop, at this moment is by far the best method to produce carvone, and improvement of carvone yields would further strengthen the competitive edge of agricultural production.

To enable improvement of carvone yields, we have studied many factors that may be of influence. This chapter will give an overview of the knowledge we have collected and of the available literature about the regulation of accumulation and composition of caraway essential oil. Regulating factors include environmental conditions and agronomic factors but also the (enzymatic) regulation *in planta*. The possibilities to use the knowledge described in this chapter to improve carvone content and yield are discussed.

6.2. ESSENTIAL OIL FORMATION IN CARAWAY

6.2.1. Difficulties when Calculating the Essential Oil Content

One of the things that may have hampered progress in studies on essential oil content – not only in caraway – is the confusing way in which contents are determined and expressed. The essential oil can be extracted using steam or hydro distillation in several different set-ups (Stahl 1953, Anonymous 1958, Luckner 1966, Stahl and Schild 1981), using whole or crushed seeds. In these distillation set-ups, the amount of oil is usually determined volumetrically which necessitates rather large seed samples to enable an exact reading. The carvone content in the essential oil is determined using a refractometer or gas chromatography. Seed samples of as little as 10 mg may be extracted using an elegant steam distillation technique developed by Krüger (1993) or solvent extraction (e.g. Bouwmeester *et al.* 1995a, Krüger and Zeiger 1993). In these latter methods, limonene and carvone are quantitated using gas chromatography. Finally, for screening of large numbers of samples (2–5 g) Near Infrared Spectroscopy (NIR) proved to be a convenient method (Toxopeus and Bouwmeester 1993).

Seed essential oil contents are usually expressed as a percentage of seed (air) dry weight, sometimes as a volume/weight sometimes as a weight/weight percentage. Using a percentage of the seed weight implies that essential oil content is determined not only by the amount of essential oil but also by the remainder of the seed. This is illustrated by data from Bouwmeester et al. (1995a) shown in Table 1. Shading of a caraway crop after flowering had a significant positive effect on seed essential oil content (Table 1). This is an unexpected result that could not easily be explained. However, the effect was entirely due to a decrease in seed weight and not to an increase in essential oil formation: the amount of essential oil per seed was not affected. To increase our understanding of the regulation of essential oil formation and have genuine progress in the improvement of essential oil content and yield it is vital to be able to draw the right conclusions (does a factor affect essential oil formation, yes or no). The conclusion that a certain factor stimulates essential oil formation when it only increases essential oil content – which may have been caused by decreased seed filling (see Table 1) – could direct our efforts in the wrong direction. We therefore strongly recommend to use absolute amounts of essential oil per seed, which can easily be calculated from essential oil contents and mean individual seed weights, in addition to the traditional essential oil content. More or less the same holds for the quality of caraway essential oil which is usually expressed as the percentage carvone in the oil:

$100 \times$ carvone/total essential oil (for 95–99% consisting of carvone and limonene).

As in the latter case carvone is both above and below the division sign, the variation in carvone content is weakened (Bouwmeester et al. 1995a). The use of a carvone/limonene ratio much clearer shows the changes in the ratio between the two compounds and

Table 1 Effect of shading from the End of flowering until crop maturity on seed essential oil content and seed weight[a]

Light reduction (%)	Essential oil[b] (% of dry weight)	Seed weight[c] (mg)	Essential oil (μg/seed)
0	1.78	3.01	53.3
19	1.74	2.94	50.8
39	1.95	2.96	56.8
69	1.82	2.88	51.9
90	1.90	2.75	51.9
P_{linear}[d]	0.018	<0.001	0.632

[a]Data from Bouwmeester et al. (1995a).
[b]Essential oil contents were determined using solvent extraction and gas chromatography and are expressed as a percentage of seed dry weight or as the absolute amount of oil in one seed (calculated from essential oil percentage and seed dry weight).
[c]Data are means of four replicates.
[d]P-values indicate the significance level for a linear relation with the percentage light reduction.

some authors have indeed started to use this ratio rather than the carvone percentage (Bouwmeester *et al.* 1995a, Galambosi and Peura 1996). Carvone content can – like for total essential oil – also be expressed as a percentage of total seed mass. Finally, carvone percentages or carvone/limonene ratios of essential oil samples obtained using (steam) distillation should be treated with care. Koedam (1982a) reported that extraction of limonene from dill seed is less efficient than of carvone due to the higher polarity of the latter. This may result in essential oils enriched in carvone when distillation is not completed (Bouwmeester *et al.* 1995a,b). Nevertheless, as much literature data as possible are

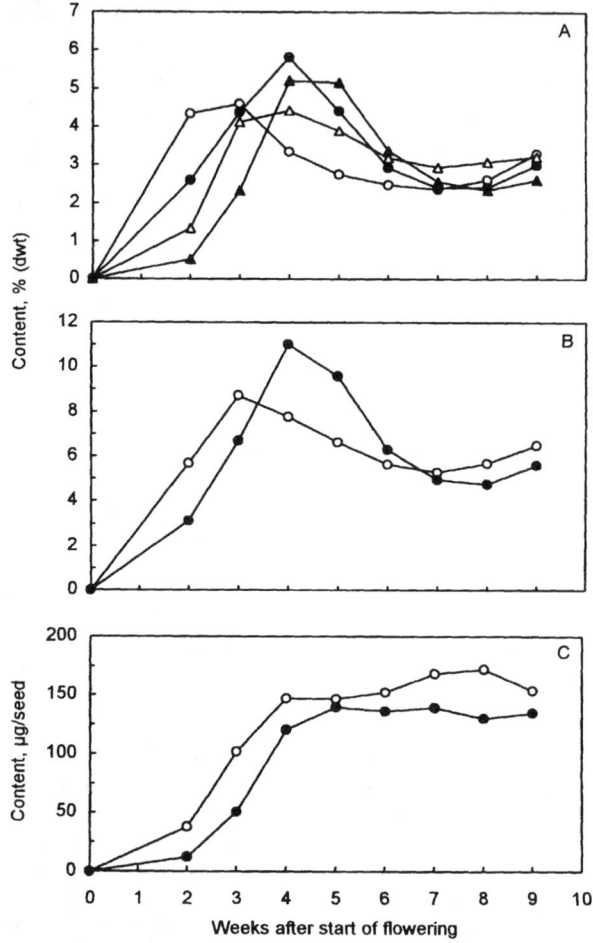

Figure 2 Developmental changes in seed contents of limonene, carvone and total essential oil (limonene + carvone) for biennial caraway. (A) Relative content of limonene (o) and carvone (Δ) in first (open symbols) and second order umbels (closed symbols). (B) Relative essential oil content of first (open symbols) and second order umbels (closed symbols). (C) Absolute amount of essential oil per seed, symbols as for B. From Bouwmeester *et al.* (1995)

used to illustrate this chapter. It can therefore not be excluded that these data also refer to essential oil and carvone percentages, some of solvent extracted, others of distilled caraway seeds. This has to be kept in mind wherever literature data are interpreted.

6.2.2. Essential Oil Accumulation During Seed Development

The accumulation pattern of limonene and carvone during development of biennial caraway was first studied by Von Schantz and Ek (1971). For annual caraway the accumulation patterns of limonene and carvone are similar (Bouwmeester *et al.* 1995a). In both caraway forms, accumulation of limonene starts earlier than of carvone, and essential oil formation in lower-order umbels precedes that in higher-order umbels (Figure 2A). Essential oil content reaches a maximum of 9–11% of seed dry weight at about 3–4 weeks after the onset of flowering and then decreases to reach a stable level of about 5–6% several weeks before the seeds mature (Figure 2B). Essential oil content of second order umbels was slightly lower than of first order umbels. In a series of other experiments in a number of different years, essential oil contents usually (but not always) decreased with umbel order (Bouwmeester *et al.* 1995a). The amount of essential oil per seed and the carvone/limonene ratio always decreased with umbel order. When the essential oil contents from Figure 2B are expressed in μg/seed (Figure 2C) it is clear that essential oil accumulation continues up to 4–5 weeks after the onset of flowering and then stops. It is also clear that the decrease in essential oil content after 4 weeks in Figures 2A, B is not caused by the disappearance of essential oil. In stead it is caused by 'dilution' of the essential oil because the seeds continue to grow. Others did report a decrease in the essential oil amount per seed at the end of fruit ripening (Luyendijk 1956, Von Schantz and Ek 1971) and also from the data of Zijlstra (1940) it can be deduced that in four out of six years the amount of oil per seed decreased with a progressively later harvest date (also see the paragraph Temperature and wind).

6.3. FACTORS AFFECTING CARAWAY SEED ESSENTIAL OIL CONTENT

An indication of the factors that may influence essential oil content of caraway can be obtained from data collected by farmers and data published in literature. Figure 3 shows data of essential oil contents of farmers in a relatively small area in the north of The Netherlands. The data show that there are large variations in essential oil contents both between farmers as well as between years. Although above we have cautioned against uncritical interpretation of essential oil contents, the data in Figure 3 are confirmed by results of others. The annual fluctuations are confirmed a.o. by Zijlstra (1940) who reported essential oil contents ranging from 1.3 to 5.7% or from 51 to 197 μg/seed. The variations between farmers are confirmed by a gas chromatographic analysis of the 1992 seed samples shown in Figure 3. Using the mean individual seed weights we could calculate essential oil amounts per seed and showed that there were indeed large variations between farmers, ranging from 45 to 144 μg/seed. The variations in essential oil contents between farmers and between years imply that there is a large array of factors that may influence essential oil content. Variations between farmers

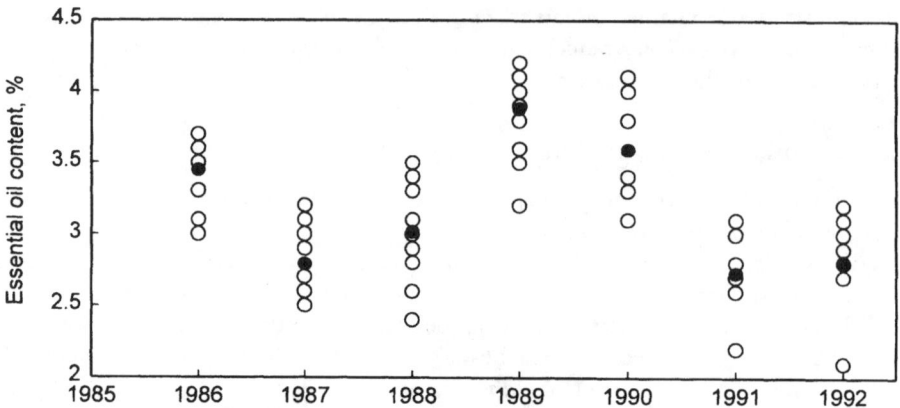

Figure 3 Essential oil contents of seed samples of farmers from the northern Netherlands over a period of seven years. Closed symbols represent yearly average essential oil contents. Essential oil contents were determined by steam distillation in a commercial laboratory. From Bouwmeester and Meijer (1994)

must be caused by differences in agricultural practices, soil properties etc. Annual variations are due to fluctuations in environmental conditions (=weather). Although there may be more factors involved I will restrict myself to the most important ones and/or those that have been investigated:

1. Agricultural practices:
   (a) Variety
   (b) Soil properties and fertilisation
   (c) Plant density
   (d) Harvest
2. Environmental conditions:
   (a) Temperature
   (b) Wind
   (c) Light intensity

An overview of what is known about these factors in literature is given in the next paragraphs.

6.3.1. Agricultural Practices

Two types of caraway are known, annual and biennial caraway, each with their own commercial varieties (see Chapter 7). The annual form is grown essentially as other annual seed crops, i.e. sown in spring and harvested in autumn. The biennial form is also sown in spring, in The Netherlands usually together with a cover crop such as peas or spring barley. In other countries, a cover crop is mostly not used. After the cover crop is harvested in August/September the caraway continues to develop a taproot until winter starts. The plants are vernalised during winter, the crop flowers in May/June and is harvested in July.

6.3.1.1. Variety

The first distinction which can be made between caraway varieties is whether they are annual or biennial. Concerning essential oil content, there is a clear distinction between these two with about 3% for annual and 4% for biennial caraway (Bouwmeester et al. 1995b). Within one of these two forms, the variation between varieties seems to be much smaller (Zijlstra 1915, Toxopeus and Lubberts 1994, Bouwmeester et al. 1995b). An illustrative experiment about the effect of the variety on essential oil content was done by Zijlstra (1916). He collected a number of seed lots from different origins including some harvested from wild caraway populations, determined essential oil contents and found large differences (Table 2). He then grew all ten 'varieties' in five replicates on one experimental field, and determined seed yield and essential oil content. Although there were large differences in seed yield (the wild populations had low seed yields), the differences in essential oil content had strongly diminished (Table 2). This indicates that differences in essential oil content are not or hardly affected by the variety but much more so by environmental conditions. Similar results were obtained by Toxopeus and Lubberts (1994) (also see Chapter 7). Ten commercial biennial varieties were compared at four different locations. Although there were two varieties with a significantly higher carvone content at one of the four locations, there were no consistent

Table 2 Essential oil contents of caraway seed samples of different origins before and after culture in The Netherlands[a]

Seed sample	After collection		After culture in The Netherlands	
	Essential oil content (%)	Deviation from mean (%)	Essential oil content (%)	Deviation from mean (%)
Moscow (probably wild)	3.48	−29	4.49	−2
Örebro, Sweden (probably wild)	6.09	23	4.65	2
Hamar, Norway (wild)	6.79	38	4.15	−9
Bamberg, Germany (wild?)	4.26	−14	4.73	4
Szénafüvek, Siebengebirge (wild)	4.90	−1	4.96	9
Württemberg, Germany (wild)	7.17	45	5.28	16
Eastern Norway (wild)	5.53	12	4.55	0
Originally Swiss, grown in The Netherlands	3.57	−28	4.19	−8
Kasan, Russia (wild)	3.36	−32	4.11	−10
Dutch (commercial)	4.17	−15	4.53	−1
Mean essential oil content	4.93		4.56	
Sum of squared deviations from mean content		7247		590

[a]Data after Zijlstra (1916).

significant differences between the varieties. However, there were clear differences in carvone content between locations with the light sandy soil giving an average carvone content of 1.6%, the sea clay 2.1% and the sandy loam and river clay 2.4 and 2.5%, respectively. No clear explanation for these differences was given but they may have been caused by the soil properties or other differences between the locations (Toxopeus and Lubberts 1994).

6.3.1.2. Soil Properties and Fertilisation

For several essential oil producing species authors have reported a negative effect of nitrogen on essential oil contents (e.g. Mihaliak and Lincoln 1985, Ross and Sombrero 1991). One of the explanations for this phenomenon may be that due to the limitation of plant growth by a lack of nitrogen, the excess carbon may be used to produce the high-energy, non-nitrogen containing terpenes (which may also protect against herbivores the little vegetative matter produced) (Baas 1989). The effects of fertilisation on caraway essential oil content reported in literature are difficult to interpret. Already in the first decades of this century it was hypothesised that fertilisation increases caraway seed yield but decreases essential oil content (Dafert and Scholz 1927/28, Potlog 1938). However, Boshart (1942) showed that this effect varied from year to year: sometimes fertilisation indeed decreased, sometimes, however, it increased essential oil content. Kordana et al. (1983) varied NPK fertilisation in a pot experiment and although they found large effects on seed yields, essential oil contents were not affected. Both Van Roon (1959) and Nordestgaard (1986) report a slightly negative relationship between nitrogen rate and caraway essential oil content. For the closely related dill, Bali et al. (1992) report no effect of nitrogen rate on essential oil content or quality, whereas Wander and Bouwmeester (1997) report a significant negative correlation between nitrogen rate and carvone content, with 1.75% for 0 kg N/ha and 1.59% for 120 kg N/ha. In the latter experiment, also the amount of carvone per seed decreased with increasing nitrogen rate, excluding the possibility that nitrogen affected essential oil content through an effect on seed filling. Interestingly, there was a highly significant negative correlation between the amounts of nitrogen and carvone per seed.

6.3.1.3. Plant Density

Zijlstra (1916) hypothesised that differences in essential oil content between two caraway production areas in The Netherlands were caused by differences in plant density. He assumed that with a lower plant density, assimilate availability would be better which should lead to a higher essential oil content. He tested his hypothesis in a field trial with four plant densities and indeed found a negative correlation between plant density and essential oil and carvone content (Table 3). Hegnauer and Flück (1949) mentioned the results of Zijlstra to explain the high essential oil contents usually reported for wild caraway plants (see Table 2), which usually grow as single plants and thus at a very low plant density. In an experiment with annual caraway we found a significant negative linear relation between plant density (20, 80 and 140 plants/m^2) and seed essential oil content or amount of essential oil per seed (Bouwmeester and Loman,

Table 3 Effect of sowing distance on caraway seed
essential oil and carvone content[a]

Distance between rows (cm)	Density within row	Essential oil content (%)	Carvone content (%)
30	high	4.89	2.52
30	low	5.14	2.67
60	high	5.03	2.67
60	low	5.44	2.88

[a]Data after Zijlstra (1916).

unpublished results). However, results of other studies are less consistent: Hornok and Csáki (1982) found slightly lower (although apparently not significant) essential oil contents at both sub- and supra-optimal plant densities, and Wander (pers. comm.) found no significant correlation between plant density and carvone content.

6.3.1.4. Harvest

Several authors have reported an effect of harvest time on essential oil and/or carvone content. Zijlstra (1940) performed a series of trials over a period of eight years in which he harvested on several dates before, at and after the optimal harvest time. In all eight years essential oil content decreased with harvest date. Toxopeus and Lubberts (1994) reported a significant decrease from 2.4 or 2.5% to 2.2% in carvone content, but an increase from 2.0 to 2.1% or no change in limonene content. When harvest was postponed for one week, Wander (pers. comm.) also found a decrease in essential oil content of only 0.1%. All these data are difficult to interpret as contents were expressed as a percentage of seed weight. If seed filling continues during the last stages of seed ripening, and essential oil formation does not, this may indeed lead to a decrease in essential oil content. Also see above the paragraph Essential Oil Accumulation During Seed Development.

The harvesting method, however, clearly affects caraway (and dill) essential oil content. Essential oil content of hand-harvested caraway seeds was 5.9% and of seeds threshed at 450 rpm 5.5% (Wander and Bouwmeester, in prep.). Increasing the threshing velocity up to 1080 rpm gradually further decreased essential oil content to 5.2%. There was a significant negative correlation between threshing velocity and essential oil content. The same significant negative correlation was found for dill with the negative effect of high threshing velocities being even higher than for caraway (4.4% at 1200 rpm compared with 5.3% for hand-harvested seeds).

6.3.2. Environmental Conditions

6.3.2.1. Temperature and Wind

Hegnauer and Flück (1949) hypothesised that because of the moist, maritime climate the Dutch and Scandinavian caraway were of superior quality compared with that from countries like Morocco, Iran (Persia), India and Russia grown under hot and dry conditions. The authors supposed that also more essential oil evaporated from the seeds in the latter countries due to less careful handling and drying. Now we know that in African and Asian countries normally annual caraway is grown which has a lower essential oil content also when it is grown in The Netherlands or Scandinavia. Guenther (1950) and Embong *et al.* (1977) hypothesised that carvone content was increased more than limonene in dry, sunny weather, either due to stimulation of photosynthesis which would favour formation of carvone more than formation of limonene or to increased volatilisation of the more volatile limonene and thus leading to an altered ratio (Guenther 1950). In line with the latter hypothesis, Wagner (1991) suggested that differences in the volatility of mono- and sesquiterpenes may be the reason that sesquiterpene contents in essential oil plants generally are higher than monoterpene levels.

To my knowledge, there has been no research on the effect of temperature on the process of essential oil accumulation in caraway. However, several authors have suggested that volatilisation of limonene and carvone from caraway seeds may occur (Guenther 1950, Luyendijk 1956, Von Schantz and Ek 1971) and it may be expected that temperature but also wind can affect this process. For a 5-year period Toxopeus and Bouwmeester (1993) reported a negative correlation between the average wind velocity from May 21 until July 15 and essential oil content of seed lots of farmers. This may support a role of volatilisation in determining essential oil content. Bouwmeester *et al.* (1995a) have measured the volatilisation of limonene and carvone from seed-bearing umbels of caraway and estimated the total loss of these compounds due to volatilisation to be about 2% for limonene and 1% for carvone. As these losses were measured under laboratory conditions (at room temperature and without strong wind), it can not be ruled out that the losses in the field may be higher. Nevertheless, we have never seen a decrease in the absolute amount of essential oil per seed (see for example Figure 2c, Bouwmeester and Kuijpers 1993, Bouwmeester *et al.* 1995a) and therefore our results contradict the results of Zijlstra (1940) and Von Schantz and Ek (1971) both reporting losses of essential oil during seed ripening, the latter even as high as 20 and 30% for limonene and carvone, respectively. For our experiments we have used non-shattering (annual and biennial) varieties. Both Zijlstra and Von Schantz and Ek report that seed shattering occurred in their experiments and it may well be that during plant maturation the seeds which were sampled were of progressively higher umbel orders (as these ripen later than the lower-order umbels). As mentioned above, there is a negative correlation between umbel order and the total amount of essential oil per seed. Moreover, the stronger decrease in the amount of carvone than in limonene reported by Von Schantz and Ek contradicts a loss through volatilisation (limonene is much more volatile than carvone) and would confirm sampling of higher-order seeds as it is in line with the decrease in carvone/limonene ratio with an increase in umbel order mentioned above. This may also explain the negative effect of wind on essential oil content reported by Bouwmeester and Toxopeus (1993). Strong wind may have increased seed shedding of the shattering varieties grown by the farmers that supplied the data on essential oil contents, particularly of the earlier ripening lower-order umbels and as the higher-order umbels mostly have the lowest essential oil content the seed shedding may have reduced the average essential oil content of the crop.

6.3.2.2. Light Intensity

Zijlstra (1915) and Boshart (1926) postulated that essential oil contents of caraway are increased by dry and sunny weather. For caraway but also other essential oil accumulating species, several authors have suggested a positive relationship between substrate (= carbohydrates) supply and monoterpene content (Guenther 1950, Burbott and Loomis 1967, Embong 1977, Hälvä et al. 1992). For caraway Bouwmeester and Kuijpers (1993) reported a positive correlation between carbohydrate content and essential oil accumulation. Field experiments in which plots of caraway were shaded at different levels showed that there was a highly significant correlation between the light intensity during flowering and essential oil content or the amount of essential oil per seed. This is in line with the data on essential oil contents that were obtained from farmers. For a 5-year period there was a positive correlation between the cumulative global radiation from March 1 to July 1 and essential oil content (Toxopeus and Bouwmeester 1993). Interestingly, in the shading experiments by Bouwmeester et al. (1995a) although essential oil content decreased with shading level, limonene content was not affected, whereas carvone content strongly decreased. As a consequence, the carvone/limonene ratio decreased with a decrease in light intensity. The important role of assimilate availability in determining essential oil content and composition is further supported by in vitro experiments in which umbels of caraway were cultured on liquid media containing a range of sucrose concentrations. An increase in the sucrose concentration up to 4% stimulated essential oil formation and more than doubled the carvone/limonene ratio (Table 4). A further increase in sucrose concentration negatively affected essential oil formation, possibly due to osmotic side effects. The positive effect of assimilate availability on caraway essential oil content is in line with results of others (Burbott and Loomis 1967, Hälvä et al. 1992). Gershenzon and Croteau (1990) also mentioned assimilate availability as a likely regulating factor of terpene biosynthesis. Not only because

Table 4   Effect of sucrose on caraway seed limonene and carvone contents[a,b]

| Sucrose (%) | Content (% of seed dry weight)[c] | | | Carvone/limonene ratio |
	Limonene	Carvone	Total essential oil	
1	3.36	1.31	4.67	0.39
2	4.46	2.67	7.13	0.60
4	3.76	3.41	7.17	0.91
6	3.77	2.81	6.57	0.75
8	3.12	2.43	5.55	0.78

[a]Data from Bouwmeester and Meijer (1994); courtesy of Iris Kappers.

[b]About 1 week after pollination umbels were detached from caraway plants and cultured on liquid media with different sucrose concentrations for one week.

[c]Limonene and carvone contents were determined using solvent extraction and gas chromatography.

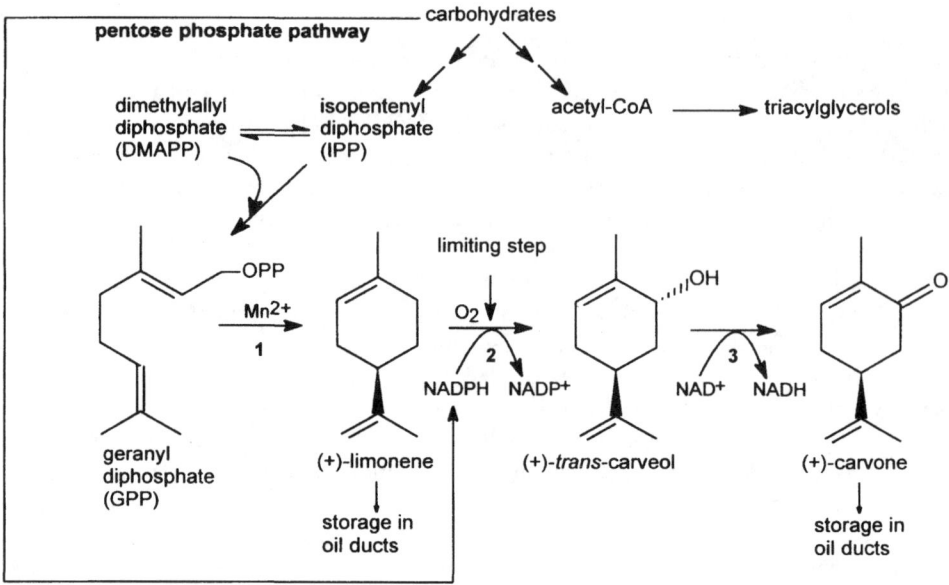

Figure 4 Biosynthetic pathway of limonene and carvone in caraway. Enzymes involved in the final three steps are: (1) (+)-limonene synthase, (2) (+)-limonene hydroxylase and (3) (+)-*trans*-carveol dehydrogenase

the basic precursors of the terpenes must be formed from carbohydrates (Figure 4), but also because energy is required for the generation of cofactors such as ATP and NADPH. Indeed the preferential stimulation of carvone accumulation by increased assimilate supply supports an important role of cofactor availability, in this case NADPH (Figure 4).

6.3.3. Conclusions

Light intensity, through its effect on assimilate availability, seems to be the most important environmental factor determining essential oil content and quality. The data available in literature all support the hypothesis of Guenther (1950) that sunny weather increases essential oil content and enhances carvone content more than limonene. This is not due to increased volatilisation of limonene under these conditions but to enhanced assimilate availability, which stimulates the formation of carvone (also see Figure 4). This conclusion is in line with the effect of plant density (Table 3). A decrease in plant density leads to a higher irradiation of the plant and thus to improved assimilate availability. The other agricultural factor that affects essential oil content is the harvest, solely because damage to the essential oil ducts leads to evaporation of essential oil.

6.4. REGULATION *IN PLANTA*

6.4.1. Introduction

Sandermann and Bruns (1964) hypothesised that in dill seed, which also contains (+)-carvone and (+)-limonene as major compounds, limonene is the precursor in the biosynthesis of carvone. Several authors have suggested that the increase in carvone content (as a percentage of seed weight) at the expense of limonene content supports this hypothesis. However, after measuring the changes in the absolute amounts of limonene and carvone in caraway seeds and using radio-labelling experiments, Von Schantz and Ek (1971) and Von Schantz and Huhtikangas (1971) showed that once secreted in the essential oil ducts, limonene is no longer available as a precursor for carvone biosynthesis. Gershenzon *et al.* (1989) described the biosynthesis of (−)-limonene and (−)-carvone in *Mentha spicata* and Bouwmeester *et al.* (1998) showed that biosynthesis of (+)-limonene and (+)-carvone in caraway occurs analogously to the production of the (−)-enantiomers in *Mentha spicata* (Figure 4). That is, a limonene synthase (1) cyclises geranyl diphosphate (GPP) to (+)-limonene, which is then either stored in the essential oil ducts or hydroxylated to (+)-*trans*-carveol by a cytochrome P-450 dependent limonene-6-hydroxylase (2). Subsequently a carveol dehydrogenase (3) oxidises *trans*-carveol to (+)-carvone, which is then also stored in the essential oil ducts.

As shown in Figure 2, accumulation of limonene and carvone in the seeds is a developmentally regulated process. In the early stages of fruit development, there is a rapid accumulation of limonene and in later stages carvone accumulation predominates, such that when the seeds mature, carvone and limonene contents are approximately equal. Although it has been suggested that carvone and limonene accumulation continue until seed maturity (Von Schantz and Ek 1971, Von Schantz and Huhtikangas 1971), Figure 2C shows that carvone and limonene accumulation cease several weeks before seed maturity.

6.4.2. Changes in Enzyme Activities

To explain the accumulation pattern of limonene and carvone in caraway seeds we collected a series of nine different developmental stages and measured (i) the contents of limonene, carvone and triacylglycerols and (ii) the activities of the enzymes limonene synthase, limonene-6-hydroxylase and *trans*-carveol dehydrogenase (Figure 4) (Bouwmeester *et al.*, 1998). The accumulation pattern of limonene and carvone shown in Figure 5A agrees well with other data (e.g. Figure 2), with the accumulation of limonene preceding that of carvone by about 5–10 days. The accumulation of fatty acids started several weeks later. The activities of the three enzymes measured varied considerably with development (Figure 5B) and showed a close correlation with the accumulation of limonene and carvone (Figure 5A). In young stages, limonene synthase and *trans*-carveol dehydrogenase were active, but limonene-6-hydroxylase was not. Thus, in this period only limonene was produced. In slightly older fruits the appearance of limonene hydroxylase activity coincided with the onset of carvone accumulation. From about two weeks after pollination, the enzyme activities started to decrease until almost

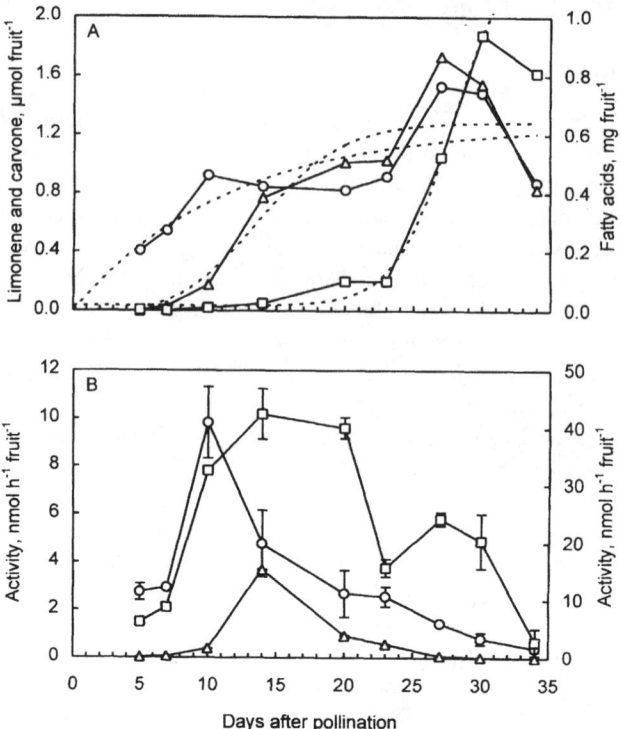

Figure 5 Developmental changes in the contents of limonene (o), carvone (Δ) and fatty acids (seed oil) (□) in caraway fruits (A) and activities of limonene synthase (o), limonene-6-hydroxylase (Δ) (both on the left y-axis) and *trans*-carveol dehydrogenase (o) (right y-axis) (B). Dotted lines in A represent fitted values

zero in the oldest stages. Although from day 20 onwards the three enzymes were still active *in vitro*, virtually no accumulation of limonene and carvone was observed *in planta*. This may be due to a lack of substrate because of the activity of competing pathways such as seed oil (triacylglycerol) formation (Figure 4). The activity of limonene-6-hydroxylase from Figure 5B is shown again in Figure 6 together with the carvone accumulation rate as deduced from Figure 5A (slope of the fitted curve). There is a close correlation between limonene-6-hydroxylase activity measured *in vitro* and carvone accumulation rate *in planta*.

6.4.3. Conclusions

The biosynthesis of (+)-limonene and (+)-carvone in caraway occurs analogously to (−)-limonene and (−)-carvone formation in *Mentha spicata* (Gershenzon *et al.* 1989). There is a close correlation between the developmental changes in the activities of the enzymes involved in the biosynthesis of limonene and carvone and the accumulation of these two compounds. We have shown that during the first weeks of fruit development,

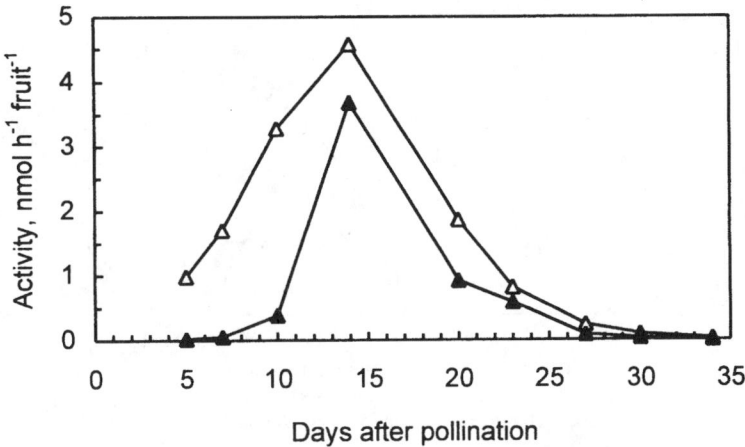

Figure 6 Developmental changes in the calculated carvone accumulation rate (open triangles; slope of fitted curve in Figure 5A) and measured limonene-6-hydroxylase activity (closed triangles; data from Figure 5B)

the activity of limonene-6-hydroxylase is the rate-determining step in the biosynthesis of carvone. After this period, formation of limonene and carvone cease possibly because of a lack of substrate.

6.5. CONCLUSIONS AND PROSPECTS FOR IMPROVEMENT

This chapter shows that the accumulation of caraway essential oil and its composition are determined by several factors. These factors are:

1. Agricultural practices: harvest and plant density, and to a lesser extent nitrogen fertilisation
2. Environmental conditions: light intensity (= assimilate availability)
3. Endogenous factors: the activity of the enzyme limonene-6-hydroxylase determines the accumulation of carvone

How can this knowledge be exploited to improve caraway seed essential oil content and quality?

1. Careful harvesting, optimisation of plant density and nitrogen fertilisation. Of these three, careful harvesting and post-harvest handling are relatively easy to realise with good improvements of yield being possible. Optimisation of plant density and nitrogen fertilisation are more difficult as both factors also affect seed yield, to a certain extent in opposite direction (an increase in nitrogen will increase seed yield but decrease essential oil content).
2. As under 1: optimisation of plant density; breeding for improved light penetration into the crop, e.g. by breeding for absence of petals or reduction in petal size; breeding for

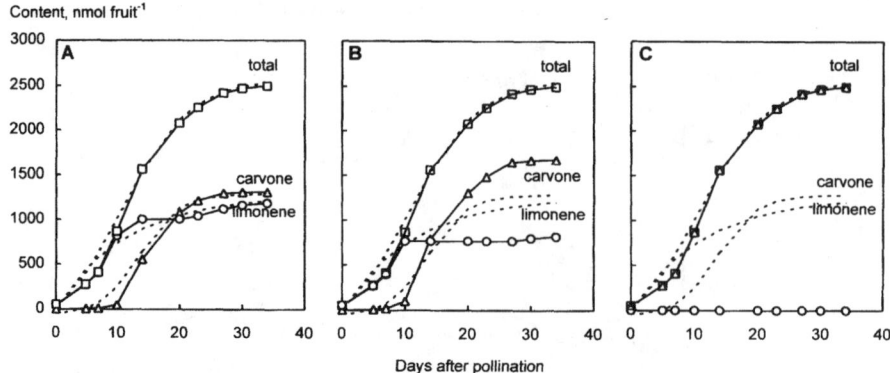

Figure 7 Developmental changes in limonene, carvone and total essential oil contents determined by gas chromatography (broken lines; data from Figure 5A) and calculated using a regression model with the enzyme activities from Figure 5B: (o) limonene, (Δ) carvone and (□) total essential oil. The activity of limonene-6-hydroxylase was corrected to account for the loss of activity during extraction. Towards maturation an increasing substrate competition was assumed to occur. (A) model describing the present situation, (B) using doubled limonene hydroxylase activity and (C) using a limonene-6-hydroxylase activity continuously at the maximum level of 14 DAP (Figure 6)

    improved carbon allocation early during seed development to improve assimilate availability during essential oil accumulation.

3. Increasing the activity of limonene hydroxylase, decreasing competition for substrate. Competition for substrate seems to limit essential oil accumulation especially during the later stages of development. Biochemical (e.g. using inhibitors of fatty acid biosynthesis) and molecular research (inhibition of branchpoint enzymes e.g. of fatty acid biosynthesis) should show the potential benefit to essential oil accumulation of a decrease in substrate competition. We have explored the potential of changes in limonene-6-hydroxylase activity using a mathematical model (Figure 7). The *in vitro* enzyme activities determined in the different developmental stages from Figure 5B were used to make a model describing the accumulation of limonene and carvone depending on developmental stage. With this model we then explored the effects of changes in the activities of the individual enzymes. Figures 7B and C show that carvone content could more than double if the activity of limonene-6-hydroxylase could be enhanced. In the future we hope to substantiate this hypothesis by isolating the gene encoding the cytochrome P-450 enzyme catalysing limonene-6-hydroxylation and overexpressing it in caraway fruits to see whether indeed the carbon flux through this terpenoid biosynthesis pathway is directed more to carvone.

ACKNOWLEDGEMENTS

Much of the work described in this Chapter was enabled by financial support of the Dutch Ministries of Agriculture, Nature Management and Fisheries (LNV) and Economic Affairs (EZ).

The author wishes to thank Jacques Davies, Iris Kappers, Eltje Loman, Martin de Rooij, Herman Smid and Maurice Konings for their contributions to the experiments described here, Jonathan Gershenzon and Rodney Croteau for their help with the enzymatic work, Willem Meijer and Kasper Hamster for their critical discussions and Hans Helsper and Bob Veen for their helpful comments upon the manuscript.

REFERENCES

Anonymous (1958) *Nederlandse Pharmacopee.* sixth edition, Den Haag, The Netherlands, pp. 32–33.

Asplund, R.O. (1968) Monoterpenes: Relationship between structure and inhibition of germination. *Phytochemistry,* **7**, 1995–1997.

Baas, W.J. (1989). Secondary plant compounds, their ecological significance and consequences for the carbon budget. In H. Lambers, M.L. Cambridge, H. Konings and T.L. Pons (eds.), *Causes and consequences of variation in growth rate and productivity of higher plants,* SPB Academic Publishing bv, The Hague, The Netherlands, pp. 313–340.

Bali, A.S., Sidhu, B.S. and Randhawa, G.S. (1992) Effect of row spacing and nitrogen uptake, content and quality of dill (*Anethum graveolens*) oil. *Indian Journal of Agronomy,* **37**, 633–634.

Beveridge, J.L., Dalziel, J. and Duncan, H.J. (1981) The assessment of some volatile organic compounds as sprout suppressants for ware and seed potatoes. *Potato Research,* **24**, 61–76.

Boshart, K. (1926) Der Öl- und Carvongehalt der Früchte – der Anbau des Kümmels in den Niederlanden. *Heil- und Gewürzpflanzen,* **9**, 6–9.

Boshart, K. (1942) Über Anbau und Düngung aromatischer Pflanzen. *Heil- und Gewürzpflanzen,* **21**, 73–90.

Bouwmeester, H.J. and Kuijpers, A.-M. (1993) Relationship between assimilate supply and essential-oil accumulation in annual and biennial caraway (*Carum carvi* L.). *Journal of Essential Oil Research,* **5**, 143–152.

Bouwmeester, H.J. and Meijer, W.J.M. (1994) Yield of caraway and carvone for industrial applications. *Proc. EC-meeting The production and impact of specialist alternative crops in the rural community,* Brussels, April 1993.

Bouwmeester, H.J., Davies, J.A.R., Smid, H.G. and Welten, R.S.A. (1995a) Physiological limitations to carvone yield in caraway (*Carum carvi* L.). *Industrial Crops and Products,* **4**, 39–51.

Bouwmeester, H.J., Davies, J.A.R. and Toxopeus, H. (1995b) Enantiomeric composition of carvone, limonene and carveols in seeds of dill, and annual and biennial caraway varieties. *Journal of Agricultural and Food Chemistry,* **43**, 3057–3064.

Bouwmeester, H.J., Gershenzon, J., Konings, M.C.J.M. and Croteau, R. (1998) Biosynthesis of the monoterpenes limonene and carvone in the fruit of caraway. I. Demonstration of enzyme activities and their changes with development. *Plant Physiology,* **117**, 901–912.

Burbott, A.J. and Loomis, W.D. (1967) Effects of light and temperature on the monoterpenes of peppermint. *Plant Physiology,* **42**, 20–28.

Dafert, O. and Scholz, R. (1927/28) Düngungsversuche mit Fenchel und Kümmel. *Heil- und Gewürzpflanzen,* **10**, 146–149.

Embong, M.B., Hadziyev, D. and Molnar, S. (1977) Essential oils from spices grown in Alberta. Caraway oil (*Carum carvi*). *Canadian Journal of Plant Science,* **57**, 543–549.

Galambosi, B. and Peura, P. (1996) Agrobotanical features and oil content of wild and cultivated forms of caraway. *Journal of Essential Oil Research,* **8**, 389–397.

Gershenzon, J. and R. Croteau, (1990) Regulation of monoterpene biosynthesis in higher plants. In: G.H.N. Towers and H.A. Stafford (eds.), *Biochemistry of the mevalonic acid pathway to terpenoids, Recent Advances in Phytochemistry 24,* Plenum Press, New York, pp. 99–160.

Gershenzon, J., Maffei, M. and Croteau, R. (1989) Biochemical and histochemical localization of monoterpene biosynthesis in the glandular trichomes of spearmint (*Mentha spicata*). *Plant Physiology*, **89**, 1351–1357.

Guenther, E. (1950) Individual essential oils of the plant families *Gramineae, Lauraceae, Burseraceae, Myrtaceae, Umbelliferae* and *Geraniaceae*, In *The Essential Oils. Vol. 4*, D. van Nostrand Company, Inc., New York, pp. 573–584.

Guérin, J.C. and Réveillère, H.P. (1985) Activité antifongique d'extraits vegetaux á usage therapeutique. II. Étude de 40 extraits sur 9 souches fongiques. *Annales Pharmaceutiques Françaises*, **43**, 77–81.

Hälvä, S., Craker, L.E., Simon, J.E. and Charles, D.J. (1992) Light levels, growth and essential oil in dill (*Anethum graveolens* L.). *Journal of Herbs, Spices and Medicinal Plants*, **1**, 47–58.

Hartzell, A. (1944) Further tests on plant products for insecticidal properties. *Contributions from Boyce Thompson Institute*, **13**, 243–252.

Hegnauer, R. and Flück, H. (1949) Versuche zur Gewinnung hochwertiger Arzneipflanzen. 2. Umwelt und erntebedingte Schwankungen im Karvongehalt beim Kümmel und ihre Berücksichtigung bei Selektionsversuchen auf hohen Wirkstoffgehalt. *Pharmaceutica Acta Helvetica*, **24**, 189–196.

Hornok, L. and Csáki, G. (1982) Effect of stand density of caraway (*Carum carvi* L.). *Herba Hungarica*, **21**, 59–65.

Janssen, A.M., Luijendijk, T.J.C., Scheffer, J.J.C. and Sven, A.B. (1988) Antibacterial and antifungal activities of caraway oil. *Proc. 19th Int. Symp. Essential Oils and other Natural Substances*, Landenberghaus Greifensee, Switzerland, September 1988.

Koedam, A. (1982a) The influence of some distillation conditions on essential oil composition. In N. Margaris, A. Koedam and D. Vokou (eds.), *Aromatic plants: Basic and applied aspects*, Martinus Nijhoff Publishers, The Hague, pp. 229–236.

Koedam, A. (1982b) Antimicrobial activity of essential oils. In K.H. Kubeczka (ed.), *Ätherische Öle. Analytik, Physiologie, Zusammensetzung, Ergebnisse internationaler Arbeitstagungen in Würzburg und Groningen*, Georg Thieme Verlag, Stuttgart, pp. 232–243.

Kordana, S., Lesniewska, S. and Golcz, L. (1983) Nutritional requirements of *Carum carvi* L.. *Herba Polonica*, **19**, 27–38.

Krüger, H. (1993) Besonderheiten in der Analyse etherischer Öle von Arznei- und Gewürzpflanzen in Züchtungsprozess. *Vorträge für Pflanzenzuchtung*, **26**, 84–91.

Krüger, H. and Zeiger, B. (1993) Determination of essential oil contents in extracts of caraway. *Drogenreport*, **6**, 31–32.

Lichtenstein, E.P., Liang, T.T., Schulz, K.R., Schnoes, H.K. and Carter, G.T. (1974) Insecticidal and synergistic components isolated from dill plants. *Journal of Agricultural Food Chemistry*, **22**, 658–664.

Luckner, M. (1966) *Vorschriften für die chemische, physikalische und biologische Prüfung von Drogen*, Gustav-Fischer-Verlag, Jena, Germany.

Luyendijk, E.N. (1956) *Over de vorming van vluchtige olie in de vruchten van enkele Umbelliferen*. Ph.D. Thesis, Leiden University, The Netherlands.

Mihaliak, C.A. and Lincoln, D.E. (1985) Growth pattern and carbon allocation to volatile leaf terpenes under nitrogen-limiting conditions in *Heterotheca subaxillaris* (*Asteraceae*). *Oecologia* (*Berlin*), **66**, 423–426.

Nordestgaard, A. (1986) Seed production of caraway (*Carum carvi* L.). Seed and nitrogen rates. *Tidskrift for Planteavlsforsg*, **90**, 37–44.

Oosterhaven, J. (1995) *Different aspects of S-carvone. A natural potato sprout growth inhibitor*. Ph.D. Thesis Wageningen Agricultural University, The Netherlands.

Potlog, A.S. (1938) Der Einfluss von Kunstdünger auf den Ertrag und die Qualität des Kümmels. *Heil- und Gewürzpflanzen*, **18**, 19–21.

Ross, J.D. and Sombrero, C. (1991) Environmental control of essential oil production in Mediterranean plants. In J.B. Harborne and F.A. Thomas-Barberan (eds.), *Ecological chemistry and biochemistry of plant terpenoids, Proceedings of the Phytochemical Society of Europe 31*, Oxford Science Publications, pp. 83–94.

Sandermann, W. and Bruns, K. (1964) Biogenese von carvon in *Anethum graveolens* L. *Planta Medica*, **13**, 364–368.

Stahl, E. (1953) Eine neue Apparatur zur gravimetrischen Erfassung kleinster Mengen ätherischer Öle. *Mikrochemie*, **40**, 367–372.

Stahl, E. and Schild, W. (1981) *Pharmazeutische Biologie, 4. Drogenanalyse II: Inhaltstoffe.* G. Fischer Verlag, Germany, pp. 20–21.

Su, H.C.F. (1985) Laboratory study on effects of *Anethum graveolens* seeds on four species of stored-product insects. *Journal of Economic Entomology*, **78**, 451–453.

Su, H.C.F. (1987) Laboratory study on the long-term repellency of dill seed extract to confused flour beetles. *Journal of Entomological Science*, **22**, 70–72.

Toxopeus, H. and Bouwmeester, H.J. (1993) Improvement of caraway essential oil production in The Netherlands. *Industrial Crops and Products*, **1**, 295–301.

Toxopeus, H. and Lubberts, H.J. (1994) Effect of genotype and environment on carvone yield and yield components of winter-caraway in The Netherlands. *Industrial Crops and Products*, **3**, 37–42.

Van Roon, E. (1959) The application of divided nitrogen dressings to some seed crops. *Report nr 6, Proefstation voor de Akker- en Weidebouw*, Wageningen, The Netherlands, pp. 131.

Von Schantz, M. and Ek, B.S. (1971) Über die Bildung von Ätherischem Öl in Kümmel, *Carum carvi* L. *Scientia Pharmaceutica*, **39**, 81–101.

Von Schantz, M. and Huhtikangas, A. (1971) Über die Bildung von limonen und carvon in Kümmel, *Carum carvi. Phytochemistry*, **10**, 1787–1793.

Wagner, G.J. (1991) Secreting glandular trichomes: More than just hairs. *Plant Physiology*, **96**, 675–679.

Wander, J.G.N. and Bouwmeester, H.J. (1997) Effects of nitrogen fertilization on dill (*Anethum graveolens* L.) seed and carvone production. *Industrial Crops and Products*, **7**, 211–216.

Wander, J.G.N. and Bouwmeester, H.J. (in prep.) Effect of threshing velocity on the essential oil content of caraway and dill seeds.

Zijlstra, K. (1915). Over karwij en de aetheriese karwijolie. *Mededeelingen van de Rijks Hoogere Land-, Tuin- en Boschbouwschool*, **8**, 1–128.

Zijlstra, K. (1916) Über *Carum carvi* L. In M.W. Beyerinck, H. Heukels, J.W. Moll, E.H. Verschaffelt, H. De Vries and F.A.F.C. Went (eds.), *Receuil des Travaux Botaniques Néerlandais Vol. 13 (3,4)*, M. de Waal, The Netherlands, pp. 159–340.

Zijlstra, K. (1940) Het verband tusschen zichttijd en opbrengst van karwij. Verslag over de jaren 1935 t/m 1939. *Vereniging tot Exploitatie van Proefboerderijen in de Klei- en Zavelstreken van de Provincie Groningen*, Groningen, The Netherlands, pp. 137–141.

Zuelsdorff, N.T. and Burkholder, W.E. (1978) Toxicity and repellency of *Umbelliferae* plant compounds to the granary weevil, *Sitophilus granarius*. *Proc. North Central Branch of the Entomological Society of America. Fifty-seventh annual conference of the North Central States Entomologists*, **33**, 28.

SECTION II
BREEDING AND
CULTIVATION

7. A CENTURY OF BREEDING CARAWAY IN THE NETHERLANDS

HILLE TOXOPEUS[1] and J. HENK LUBBERTS[2]

[1]*Nassauweg 14, 6703 CH Wageningen, The Netherlands;*
[2]*Centre for Plant Breeding and Reproduction Research (CPRO-DLO) P.O. Box 16, 6700 AA Wageningen, The Netherlands*

7.1. INTRODUCTION

7.1.1. Name, Types, Origin, Use, Markets

Caraway (En), Karwij (Nl), Kümmel (G), Carvi (Fr, Sp, It), are European common names for the plant *Carum carvi* L. and its seed. What is commonly referred to as the 'seed', as in this text, is botanically a small dry fruit with a 1000-seed weight of 2–7 g. The species occurs in a biennial (*C. carvi* L. cvg Winter caraway), and an annual form (cvg Spring caraway) (Toxopeus *et al.* 1996). The former is part of natural pastures in Central Europe; Alps, Caucasus, Adriatic peninsula and the drainage areas of rivers notably Rhine and Danube (Hegi 1926). The annual type is natural in the Middle East and the eastern Mediterranean.

The seed is used in bakery products and appreciated for its pungently fresh taste upon biting it, it is sold mainly on appearance in regional and world markets.

Caraway essential oil (2–7% of weight) is (steam) distilled from the seed. The oil is used in a variety of cosmetic products. Vacuum distillation of the oil produces carvone.

The seed stores well and is easy to transport, it is an international commodity subject to international markets and pricing.

7.1.2. History of Production

Statistics in The Netherlands show that caraway was cultivated as a field crop as early as the end of the 18th century, just before crop statistics were started (Kops 1840). Caraway was first cultivated as field crop in the area around the town of Enkhuizen on the coast of the then 'Zuyderzee' inland sea in the province of Noord Holland. The variety is referred to as the 'Noord-Hollandsche' landrace (Rassenlijst 1936) which, ultimately, must have been derived from naturally occurring caraway but of this there is no documentary evidence. In the early 1900s caraway cultivation was initiated in the Oldambt region with very heavy sea derived clay in the Northeast of the northern province of Groningen. In the 1920s and -30s, the heyday of caraway production in The Netherlands, when over 20,000 ha were grown, it was cultivated in all provinces (Zijlstra 1916).

7.1.3. Varieties in The Netherlands

Between 1902 and 1908 Dr. R.J. Mansholt of the Mansholt seeds firm had selected the variety 'Mansholts karwij' from the 'Noord-Hollandsche' landrace. The former was included in 1924 in the first Dutch 'Rassenlijst', the official descriptive list of varieties of field crops (Broekema 1924). In 1930 the variety 'Volhouden', selected between 1924 and 1930 by farmer Kistemaker of Anna Paulowna c. 60 km north of Enkhuizen, was included in the Rassenlijst (1930). In 1936 the 'Noord Hollandsche' landrace is included in the Rassenlijst (1936) to protect it from extinction. It was maintained and offered for sale by the 'ZAP' a farmers' co-operative in Anna Paulowna. According to the various variety descriptions differences between the three varieties, if any, were minimal. Eventually, in 1965, the 'Noord-Hollandsche' was removed from the Rassenlijst (1965) and does not as such exist anymore. Nor does, since 1993 'Mansholts karwij' (Rassenlijst 1993). Since 'Volhouden' is maintained as a broad population, i.e. has a large effective population size, it may safely be considered to be fully representative of the two former varieties. Volhouden is maintained on the Kistemaker farm and is marketed by the ZAP.

In 1972 the variety 'Bleija', a product of 'Volhouden' and 'Mansholts' producing a non-shattering seed, appeared for the first time in the Rassenlijst (1972).

1.4. Production in The Netherlands

The Oldambt, the northern polders in Noord-Holland province and the heavy soils in Zeeuws Vlaanderen in the extreme Southwest are presently the mainstay of caraway cultivation. However, this amounts to only several hundreds of hectares in the past few years.

7.2. GENETICAL AND PHYSIOLOGICAL ASPECTS, BREEDING METHODS AND AGROTECHNOLOGY

7.2.1. Flowering and Pollination, and Inheritance of Flowering Induction

In the field the pollen is carried by wind and by a variety of flying insects (Bouwmeester 1995), reproduction is by seed. The mating system is cross-fertilising i.e. most of the flowers are fertilised with 'cross' pollen. This and the fact that even for small field plantings several hundreds of seeds are required, secures the continuation of a large effective population size (Allard 1960) of the traditional varieties. Consequently even genes with very low frequencies are kept going in these cultivars.

In the field, groups of selected plants are best isolated from other pollen sources spatially. At CPRO-DLO spatial isolations (combining different crops) were kept at a minimum distance of 50 m in all directions usually in a large field of winter-wheat sown specially for the purpose.

In the still air of the green house environment in autumn-winter-spring the convection air currents were apparently not capable of making pollen move because fruits will grow empty of seed. A combination of wind generated by ventilator, and/or bees resulted in good seed set, allowing for seed, i.e. a generation, to be produced in winter.

Flowers are complete and are organised in umbels of which the outside flowers mature first and the other follow concentrically inward. The flower is protandrous, the pollen matures 1–5 days before the pistils do (Zijlstra 1916), the higher the temperature the shorter the interval (Keulen 1988).

The procedure for controlled pollination at CPRO-DLO takes advantage of these phenomena. Bagged umbels produce some seeds when shaken from time to time, and fair amounts of selfed seeds may be produced by bagging several umbels of the same plant together. In the winter of 1993 some 40 plants thus bagged, in the greenhouse, produced several hundreds of seeds per plant without any indication of occurrence of self incompatibility. Keulen (1988) demonstrated that bagged umbels with central flowers removed will not set seed upon shaking, i.e. the last pollen produced has died before the first pistils mature.

As a final check on the method, reciprocal crosses between annual plants and Bleija biennial plants were made. Properly prepared female receptive umbels (43 of each parent) were pollinated with pollen shedding umbels of the other parent. Average seed set of individual flowers was 60%, over 2300 seeds were produced. Seeds were sown early July in pots set up in a cold frame. By the end of August pots and plants were transferred to a greenhouse at 18°C and a daylength of 14 hours. Observations on flowering continued until February. The cross biennial × annual produced 65% annual plants (out of 158); the reciprocal cross gave 62% annual plants (out of 169); the difference was not significant. Effects from selfed seed occurring, would have decreased the percentage flowering plants in the former cross, and would have caused an increase in the reciprocal. There was no such indication and therefore no evidence of occurrence of plants from selfed seed, i.e. the method of crossing gives 100% crossed seed. Out of 165 plants from purely annual seed, the check, 95% flowered (Keulen 1988).

The absence of a significant difference between the reciprocals implies the action of Mendelian genes. The simplest hypothesis to explain the segregation of annual and biennial plants as above, is one of a single locus with either incomplete dominance of the allele for annual flowering, or complete dominance but with a small frequency of the biennial, recessive allele in the population. A frequency of 0.22 is implied by the occurrence of 5% non-flowering plants in the progeny of the annual check as above. However, this would result in 78% flowering plants in a cross rather than the 65% as above. This discrepancy may be explained by experience from other greenhouse experiments which shows that the crosses would have yielded more and very late annual plants if the observations had continued for another one or two months.

7.2.2. Ecological and Agrotechnical Aspects

Farmers experience shows that plants of the biennial type can only be vernalised when the taproot is the size of a pencil or more. Research in 1987 at the former Foundation for Agricultural Plant breeding SVP (presently incorporated into CPRO-DLO) showed that such plants require a period of about 8 weeks of temperatures below 10°C (and a little light and water to keep them alive), after which they will flower when grown at normal temperatures.

The annual type is induced to flower by long days (10 hours or more), the higher the temperature the quicker flowering develops.

Cultivating the annual variety Karzo is a matter of sowing early in spring and harvesting early autumn, its seed does not shatter.

Cultivating biennial varieties is awkward mainly because only plants with pencil-thick taproots and larger can be induced to flower after the winter (vernalised). Because juvenile plant growth is slow the crop has to be sown early. The latest possible time of sowing in The Netherlands is the end of June which makes the production of a previous crop impossible. This means either no income in the year of sowing, or sowing the caraway in spring, together with a suitable companion crop for income. The latter is common practise in The Netherlands and companion crops are an important issue.

Another complication of the cultivation of winter caraway is that the seed produced in one year cannot be used to sow that years' crop. Periodicities of biennial varieties are not well adapted to prevailing agro-economical seasons.

Prior to harvesting varieties which fruit shatters when dry and ripe (see Chapter 7.2.4.), the crop is cut in swaths at the first sign of ripening fruit shattering, in early July, to dry and subsequently to be picked up and threshed. The clean break between fruit and stalk induced by the abscission layer (Figure 1) produces a very clean seed. The crop of a non-shattering variety may be combine-harvested by the end of July. Threshing and

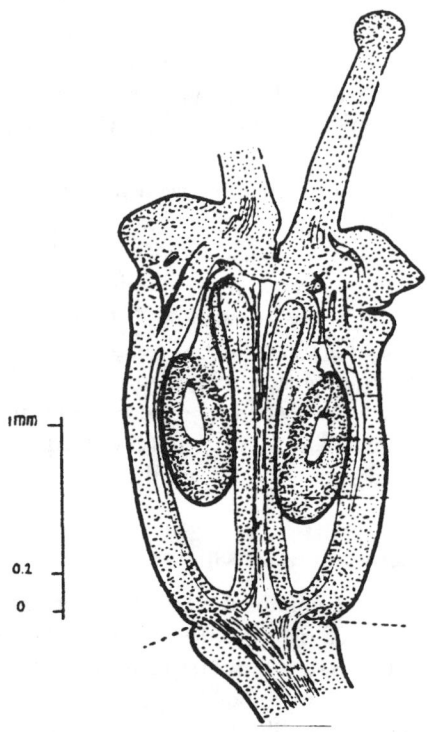

Figure 1a

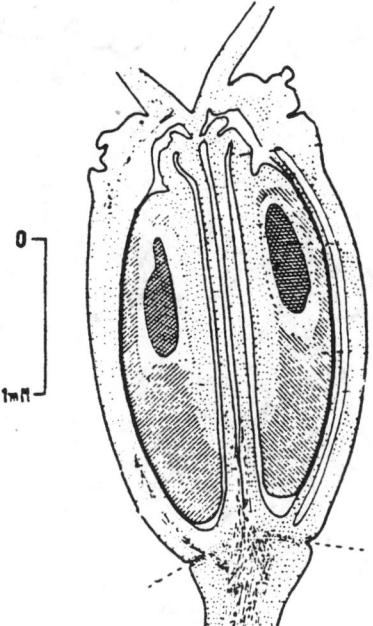

Figure 1b

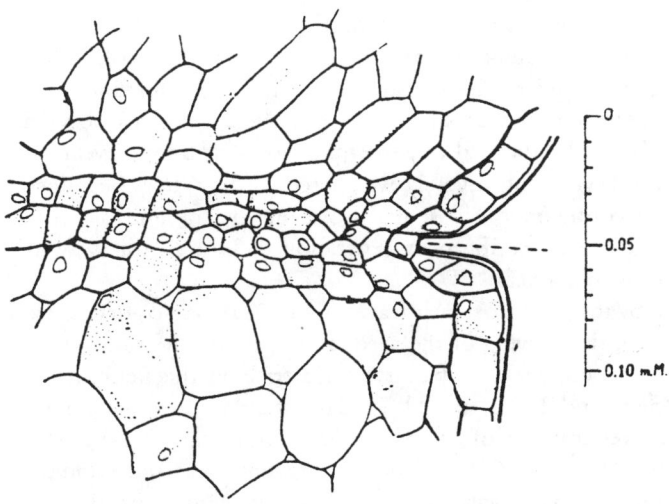

Figure 1c

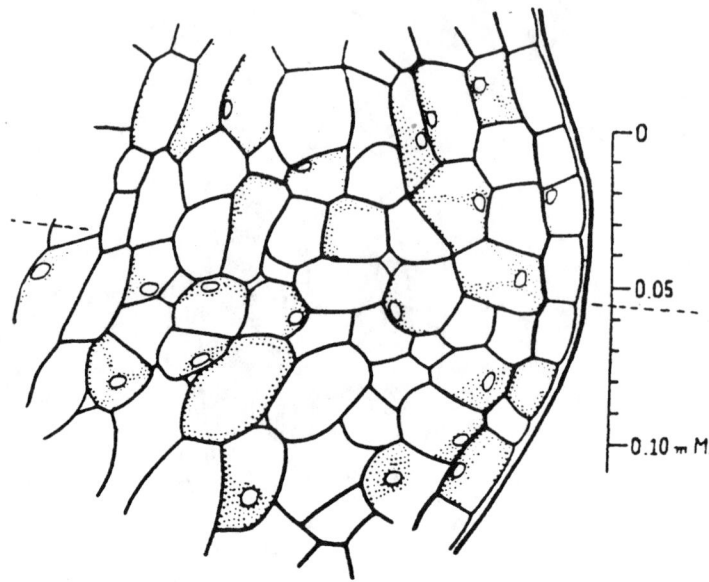

Figure 1d

Figure 1a-d   Shattering and non-shattering Caraway A comparative anatomical study of flower and fruit (After Goedewaagen and Zijlstra 1925) All cuts median, dotted line indicates position of abscission layer. (1a) Flower (double) at anthesis, shattering type; (1b) Ripening fruit (double), shattering type; (1c) Abscission layer of ripening fruit, shattering type; (1d) Location of the non existing abscission layer, non-shattering type

cleaning need more attention so as to get rid of stalks attached to seeds, a bad quality trait. The more intensive threshing may affect oil content adversely.

An important breeding method involves the use of a cold frame facility to sow seeds by the end of August. The resulting accelerated germination and juvenile plant growth make plants at the end of autumn of a sufficient size to be vernalized, and are effectively vernalized by the end of December. At that time of year plants consist of mainly roots, the foliage being severely weathered. 'Roots' may be transplanted to the field in the desired configuration in spatial isolations by the end of December or early spring, to flower and produce the next generation in July. In this way it is possible to grow a generation within a year.

Sowing by the end of August leaves about 7 weeks for drying harvested plants, threshing, cleaning, final drying of the seed, and selection, following harvesting in early July.

Methods to conduct comparative yield trials in the field are reported in some detail by Toxopeus and Lubberts (1995). The standard complete randomised block design with three replications of 45 sq.m plots comparing 18 varieties gave very satisfactory results and an overall CV of 7%. Sowing time was the middle of June, the use of a companion crop was considered unduly complicating and risky. Sowing and harvesting were carried out with appropriate (HEGE) standard field research equipment.

7.2.3. Methods of Clonal Propagation and Genetic Modification

A successful method of cloning selected plants using very young umbel frames for explants was developed by Neervoort and co-workers at the Van Hall institute, Leeuwarden (Toxopeus *et al.* 1996). Umbels with young to very young buds were cut from robust healthy plants in a promising population, and immediately transferred to a cool box. After cutting off the buds, the umbel frames were sterilised and cultured upside down on Gamborg's B5 medium supplemented with NAA and BA. The cultures were maintained for 5–6 weeks until the carvone content of the parent plant's seed had been determined. Clones of superior parent plants were selected for multiplication. Subsequently the cultures were transferred to MS10 medium for multiplication, and were grown into plantlets. These were eventually transplanted to potting mixture in the greenhouse and grown into full plants, such plants will be referred to as cuttings.

Cuttings originating from annual plants would flower immediately having probably been induced to flower while being cultured in continuous light. Cuttings of biennial genotypes were difficult to judge for sufficient size to be vernalized because they grow a somewhat diffuse ball of roots with 2–4 small taproots, instead of the firm single seedling taproot.

Since biennial cuttings will remain vegetative if not vernalised the plants can be kept going for years and may also simply be multiplied by splitting up.

Krens *et al.* (1997), at CPRO-DLO, report an efficient and stable method to produce transgenic plants from cotyledonary node explants inoculated with *Agrobacterium tumefaciens*. The percentage of explants giving one to several transgenic plants may be as high as 13%.

7.2.4. Fruit and Essential Oil

The flowers occur in pairs that grow into double fruits which separate when drying out at maturity. The traditional winter type grows fruits that shatter when dry ripe due to an abscission layer in the stalk (Figures 1a–c). The absence of the abscission layer results in the 'non-shattering' (NS) fruit (Figure 1d).

Caraway essential oil is excreted and stored in specialised vessels called vittae located in the fruit exocarp separated by a few cell layers only from the open air. The oil is thus embedded in a maternal tissue and indeed a product of maternal origin. (This is in contrast to seed oils made up of glycerol and fatty acids such as rape oil are tucked away inside the cells of seed tissues such as endosperm or cotyledons and consequently have a zygotic origin.)

The oil consists of two main components, both monoterpenes: S-(+)-carvone and S-(+)-limonene, making up from 50–60% and 35–45% of the total respectively, the remainder consisting of a mixture of compounds. Carvone in particular is biologically active and has a variety of potential uses. It is by far the most interesting component, hence the efforts to develop both potential products and markets (Toxopeus and Bouwmeester 1993).

The biochemical pathway shows that limonene is the progenitor of carvone and is converted into carvone enzymatically.

7.2.5. Methods of Chemical Analysis

The three caraway breeding research programmes conducted in Wageningen employed the following methods of analysis of oil contents of small samples.

Zijlstra (1916) developed an indirect method based on the lowering the freezing point of the solvent ethylene bromide as a result of the oil extracted from 5 g samples of finely ground caraway seeds. Six samples were required for the accurate determination of the oil content of a seed batch. Carvone was determined by refraction of a small amount of oil distilled by high temperature steam.

A steam distillation protocol using Stahl apparatus was operational at the former SVP as from the late 1960s (Krechting 1970). Air-dried samples (c. 10% MC) of 4.5–9 g of whole but ruptured seed are extracted in boiling water for 1 hour. The distilled oil is deposited floating on a water column. A sample of the oil is used for a determination of carvone content by refraction.

Analyses by Gas Chromatography (GC) became the standard from about 1980, requiring samples of less than 1 g (Oostrom and Van der Kamp 1990). Samples are very finely ground and oven dried, carvone and limonene are individually identified, and total oil content is derived at by adding the two.

Steam distillation does not extract the full oil content from samples so that values are lower than those of the other methods described. Moreover the use of air-dried samples also decreases contents. However, the data are a more accurate reflection of the results of commercial extraction which employs overheated steam. Another advantage is that small amounts of oil may be produced for further studies.

For good comparison, data on contents should not only state analytical methods, but also moisture content of samples.

An accurate indirect and non-destructive method was developed at CPRO-DLO in 1988, employing near-infra red spectral analysis (NIRS) requiring c. 2 g samples. Initially a Compscan 3000 with a spinning mirror calibrated with a series of steam distillation data was used during 1988–1992 (Toxopeus and Bouwmeester 1993). Its predicted values of samples analysed with GC correlated well (r = 0.85). In subsequent studies at RIKILT-DLO (Tusveld *et al.* 1991) a NIRS-6500 and a Brann and Luebe Infralyser 500 were used, both equipped with monochromator. Using multilinear regression analysis, the comparative study of calibration lines revealed a line based on 2 wave lengths 1626 and 1736 nm, producing data correlating with a coefficient of 0.95 with the independently analysed series (by GC) and with a Root Mean Square error of calibration (RMSC) of 0.19.

At CPRO-DLO an Infralyser 500 calibrated with GC data is in use as from 1993. The calibration line is annually adjusted with values of the new years crop.

7.3. BREEDING AND BREEDING RESEARCH

7.3.1. Zijlstra, "Zaadhoudende karwij" and Botanical Research

"Zaadhoudende Karwij" was the name of a non-shattering (NS) variety the initial plants of which had been found in the vicinity of Wageningen in 1915 (Zijlstra 1916,

p. 337/8). An unspecified process of selfing and observing segregation in selfed progenies, selecting and selfing homozygous NS plants eventually yielded a 'pure' variety.

In the mean time it had become clear that the NS character showed a dominant and
simple inheritance. Goedewaagen and Zijlstra (1925) describe the NS character in detail
and visualise it in the anatomical drawing as in Figure 1. Yields of 2 t/ha and over are
reported, and Zijlstra (1940) confirms the same yield level in 1939.

The variety appeared in the first Dutch list of recommended varieties in 1924 classified as variety worth of further testing (Broekema 1924) commenting that the crop
plants are somewhat slender. The variety disappears from the list in 1927 unfortunately
without comment. The agronomic characters of the variety were apparently not up to
standard. As the underlying cause inbreeding depression is suggested in reports concerning the 'Bleija' breeding programme summarised below. Zijlstra's extensive treatise
of 1916 contains detailed descriptions and drawings of morphological and anatomical
features of plant parts with particular attention to flower and mating system, fruit and
vittae. He was particularly concerned with the carvone component of the oil which he
believed to be by far the most valuable substance of the caraway fruit. The analytical
method used is outlined in the former chapter.

With the purpose to get a better understanding of the environmental effects on
oil and carvone content, two interesting studies were conducted which results are
summarised here.

(1) Caraway seed samples obtained from different parts of Europe were analysed for
oil and carvone content. Two years later samples were analysed of the respective progenies grown in replicated plots inside a commercial field in The Netherlands (Zijlstra
1916). Table 1 shows the data. Contents of the original samples vary a great deal, from
3.4 to 7.2%. However, contents of the progeny samples produced in The Netherlands
converge to around the average content obtained in this country, i.e. between 4.0 and
4.5%. This seems to indicate that most of the variation obtained in the original samples
was due to environmental factors including seed processing. Nevertheless, the sample
from Wuertemberg Germany was also the highest when grown in The Netherlands and
may well express a heritable character. The samples from Kolozsvar are very nearly
equal and consistently high, also showing possibly a heritable trait.

(2) Some 25 years later, Zijlstra (1940) presents data of nine years of field experiments to study the effect of the date of cutting the crop in windrows on yield and seed
values of the variety "Noord-Hollandsche". At each date of cutting, duplicate plots
were cut out of a large field at the Groningen Experimental Farm on Oldambt clay, for
treatment and analysis. In the first 5 years three harvesting dates were used: 'early',
'timely' and 'late' with 3–4 day intervals. It should be kept in mind that due to the
shattering nature of the fruit, cutting could not be delayed much without loosing a lot
of seed. In the last three years two earlier dates of cutting were added. At each date of
harvesting: *seed yield, 1000-seed weight* (except for the first two years), *oil contents* of air-dry
as well as oven-dry seed, and *carvone content* of the oil were determined.

The oil content data enabled the computation of seed moisture content (MC).
In Table 2 the seed values are shown, recalculated to a standard 11% MC. The volume
of oil in 1000 seeds, '1000-s volume' was computed using 0.9 as the standard density of
the oil to convert weight to volume.

Table 1 Essential oil and carvone content of oil in samples of commercial caraway seed and of their progeny grown in The Netherlands. (Modified after Zijlstra 1916)

Origin of the seed	Essential oil	Carvone	Essential oil	Carvone
	Original sample		Sample progeny	
Moscow, Russia, wild	3.48	50.5	4.49	52.5
Kasan, Russia, wild	3.36	50.0	4.11	54.0
Oerebro, Sweden, wild	6.09	55.0	4.65	55.0
Hamar, Norway, wild	6.79	55.5	4.15	55.0
E. Norway, wild	5.53	49.5	4.55	54.5
Bamberg, Bavaria, Germany	4.26	53.0	4.73	52.0
Kolozsvar, Hungary, wild	4.90	53.0	4.96	55.0
N.E.Wuerttemberg, Germany, wild	7.17	53.0	5.28	56.6
Local NL., seed for sowing	4.17	55.5	4.53	52.5

Table 2 Effect of year, and time of cutting the crop, on yield and seed values of caraway variety 'Noord-Hollandsche' 1931–1938. (Modified after Zijlstra 1940)

Year/date of harvest	Seed yield kg/ha	1000-s weight mg	%EO	%C in EO	EO yield kg/ha	EO 1000-s volume μl
Annual effect on yield and seed values						
1931, 6/7	1450	—	4.46	51.5	65	—
1932, 5/7	1750	—	4.20	54.0	74	—
1933, 5/7	465	3452	5.35	58.6	25	205
1934, 6/7	1125	4388	1.04	60.8	12	51
1935, 4/7	1338	3000	5.42	49.8	73	181
1936, 4/7	2387	3226	4.32	46.4	103	155
1937, 5/7	870	3039	2.45	55.8	21	83
1938, 4/7	1560	3222	3.77	52.0	59	135
The effect of time of cutting the crop on yield and seed values						
1936, 22/6	1422	2159	6.53	47.9	93	157
26/6	1883	2569	5.54	49.1	104	158
30/6	2364	2973	4.58	48.0	105	151
4/7	2387	3226	4.32	46.4	103	155
8/7	1496	3387	3.54	46.3	53	133
1937, 22/6	782	2473	3.13	56.2	25	86
26/6	895	2756	2.90	56.5	26	89
30/6	893	2993	2.67	56.5	24	89
5/7	870	3039	2.45	55.8	21	83
8/7	865	3025	2.59	51.8	22	87
1938, 22/6	995	2188	6.49	50.5	62	158
26/6	1251	2412	5.09	52.3	64	136
30/6	1434	2910	4.58	52.9	66	148
5/7	1560	3222	3.77	52.4	59	135
8/7	1302	3455	3.73	52.0	49	143

EO = essential oil; C = carvone; s = seeds; (MC of seed values: 11%).

Zijlstra (1940) only briefly points to the widely fluctuating yields, the decline of oil content and the increase of 1000-seed weight as the crop is cut at a later date. The increase of fruit weight is understood to be the result of the embryo still accumulating weight in the ripening process. The possible relation with oil content is not mentioned. In Chapter 7.4 the issues of yield and seed values, and their relations are discussed.

7.3.2. Non-shattering "Bleija" and its Autotetraploid

In 1960, the former Foundation of Agricultural Plant Breeding SVP, now CPRO-DLO, started a programme to breed a productive NS variety, making a fresh start.

It was argued that there was a good chance to find NS plants in one of the traditional varieties considering their broad effective population size. Wit, conducting a search in June 1960 identified five NS plants amongst c. 2500 screened in the border of a commercial field of Volhouden. In the barn of the farm, amongst a heap of threshed plants from the previous crop harvest, two plants were found with most of the seed still attached. An accurate selection method was used: visual observation (aided by hand lens if required) of the presence or absence of the abscission layer in the stalks of flowerbuds, flowers and fruits well before pollination.

The half-sib progeny of each of these seven NS plants formed the basis of the breeding programme. At the time of fruiting the following year, 55% of a total of 369 plants turned out to be NS. This proportion does not significantly deviate from a 1:1 segregation, in line with the hypothesis of the NS character being monogenic dominant. Van Roon and Bleijenberg (1964) formally proved the hypothesis to be correct.

The low frequency of the NS allele in the population indicates that the parent plants were probably heterozygous for NS. The frequency of the NS gene in the variety Volhouden (five heterozygous NS plants in 2500) is 0.002.

In 1962 fourteen excellent plants of the (other) variety 'Mansholt's karwij' were crossed with pollen of the 50 best NS plants so as to avoid the possible ill effects of inbreeding depression. In 1965, 67 families homozygous for NS were available (Wit 1989). Selecting for earliness, uniform flowering, prolific setting of well filled fruit, and lodging resistance reduced this to 33 top families in 1968. For comparative field testing seed was bulked up of several testing combinations between families. The conclusion of several years testing was that the NS material yielded 20–30% more than the traditional S varieties as a result of losing that much less seed in the harvesting process. Twelve families were eventually selected as basis of the NS variety 'Bleija'.

An equal mixture of seed of each of the families provides basic seed for multiplication which is allowed to be continued for 2 generations. The families are kept at CPRO-DLO, the seeds firm ZAP mentioned above takes care of multiplication and marketing.

Bleija was included in the official list of recommended varieties in 1972, stating that its yield and agronomic characteristic was equal to that of the other two varieties but that the oil content of its seed was slightly lower (Rassenlijst 1972).

The research on the comparative performance of Bleija reported by Toxopeus and Lubberts (1995) confirmed these observations, quantifying in particular the losses due to the seed shattering character. If swathed and treated in the same way that the S varieties are at harvesting, Bleija's seed has the same oil content and constitution as its parent varieties. Probably as a result of combine harvesting the dry ripe crop on the field, the seed appears to loose some of its limonene component causing a slightly lower oil content. However, because carvone content is not affected, its proportion in the oil becomes higher (c. 70%) (Wander 1994).

In the late 1940s and early 60s polyploidy was reported to secure large increases of caraway essential oil content. Therefore it was decided to develop a tetraploid NS population on the basis of the Bleija families (Dijkstra and Speckmann 1980).

In a first experiment several thousands of pregerminated (Bleija) seeds were soaked in a range of colchicine solutions for different periods of time. At a concentration of 0.025% a few plants survived providing four tetraploids which plants served as model for the selection of more to come. A distinguishing trait of tetraploid plants is the rough triangular shape of the pollen grains as against the smooth oblong shape of diploid pollen. Stainability of the pollen from tetraploid plants was poor and the seed was much larger than the diploid seed.

Next the growing points of nearly 1000 seedlings with 2 or 3 leaves were treated with a solution of 0.2% colchicine for 2 days; 20% of the plants looked like being mixoploid. After selection for pollen size and big seeds, up to 50 big seeds taken from 35 C0 plants were sown to produce the next, C1 generation. Eventually 157 tetraploid plants were identified, derived from 10 of the 12 Bleija families, and were planted together in a field isolation. The 64 plants with the best pollen stainability ($>45\%$) and/or a 50% seed set were finally selected as parents for the C2 generation. Germination rate of the C2 seed was 70%, a considerable improvement on the measly rate of 27% of the C1 generation.

The C2 and C3 populations each consisted of 1000 plants c. 10% of which were selected as parents for the next generation. Comparing the two generations, pollen stainability was at the same level of 59%, and both had a seed set of 50% so since the C1 there had not been any improvement. The C3 material showed a darker green foliage and a more compact plant that was a little later in flowering and one week later to ripen than diploid plants. 1000-seed weight was 7 g, as compared to 3 g of diploid plants. Oil content (by steam distillation) was between 3 and 3.5% with 72–74% of carvone in the oil, as against $2.2 \pm 0.2\%$ oil content and $67 \pm 5\%$ carvone in the oil of Bleija seed.

Meiotic observations were difficult because of the small flowerbuds and the lack of synchronisation in meiotic events between similar sized buds. Diploid material ($2n = 20$) exclusively showed 10 bivalents. Twenty pollen mother cells from three C3 plants, at diakinesis had an average of 0.8 (0–2) univalents, 16 (12–18) bivalents, 0.4 (0–2) trivalents and 1.4 (0–4) quadrivalents. In 12% of PMCs at anaphase I, 1–2 laggards occurred. A micronucleus was observed in 8% of 177 tetrads; no inclusions were found in young pollen grains (Dijkstra and Speckmann 1980).

The C3 yielded C4 seed in bulk in July 1977, and was further multiplied to the C7 in bulk in 1984. The yield level was between 500 and 700 kg/ha, and the 1000-seed weight about twice as heavy and the oil content about 50% higher than diploid parent Bleija. The impression was that the crop domestication process had not resulted in a higher productivity (Dijkstra 1993), the partial sterility was still present.

7.3.3. Breeding for Higher Carvone Production and Diversifying the Caraway Crop

As from 1984, caraway genetic improvement received a new impetus in the context of multidisciplinary research projects to create new, industrial, crops (Toxopeus and Bouwmeester 1993) amongst which caraway. The idea was to develop and market a potato sprout retarding agent on the basis of S-(+)-carvone, as found in caraway essential oil.

The main objective of breeding research was to attempt to increase carvone content of the seed whilst at least keeping up the existing seed yield level, and in this way improve carvone productivity.

However, being confronted with the annual plant material as from 1985, a programme to develop useful, well tested annual varieties was carried out. Finally, genetic variation was introduced and evaluated.

The following breeding research and variety development programmes were conducted.

7.3.3.1 Recurrent Divergent Mass selection in the traditional, biennial 'Oldambt' landrace, RDMO, to study response to selection both for high and for low contents of oil and carvone.

7.3.3.2 Develop biennial experimental varieties with higher contents

7.3.3.3 Develop dependable spring caraway varieties

7.3.3.4 Domesticate a winter caraway type capable of being sown in autumn

7.3.3.5 Improve fertility and evaluation of the tetraploid population "SVP-4X".

7.3.3.6 Evaluate germplasm collected in The Netherlands and southern Germany.

7.3.3.1. RDMO

The basic idea was to attempt to stepwise increase frequencies of genes for **high**, respectively **low** contents of oil yet leaving (gene-frequencies of) other characters unaffected. In order to achieve this, population size was not to be narrowed down to less than 30 effective parents, theoretically causing an acceptable decrease of heterozygosity of about 1.7% in each generation (Allard 1960). Moreover, cross pollination amongst plants in a selected population was to be prolific so as to break linkages. Populations were planted as spaced plants in spatial isolations and seed production was left to the natural (cross) pollination process.

The success of the programme depended to a large extent on the availability of a non-destructive, rapid yet reliable method to determine oil and carvone content of samples produced by single field plants, i.e. 5 g or less.

The first cycle of selection started in 1985 by determining the oil and carvone contents by means of steam-distillation (the NIRS method was not yet available). The 4,5 g of seed was analysed of each of 262 plants from a commercial field sown with farmer's own seed 'Oldambt'.

The 262 plants were selected into the following five groups:

EOH, consisting of (the rest seed of) the 16 plants highest in essential oil content, and *EOL*, made up of the 8 plants lowest in oil.

COH, consisting of the 26 plants highest in carvone content in the oil, and *COL*, made up of the 4 plants lowest in carvone content in the oil.

CSH, consisting of the 13 plants with the highest (calculated) carvone content of the seed.

– The first generation:

Rest seed of the selected plants was sown in the cold frame in June 1986 producing vernalized plants in December. These were transplanted to the field, in spatial isolation, by group according criteria of selection.

The following numbers of plants flowered in each isolation in the spring of 1987: EOH1 162; EOL1 88; COH1 404; COL1 71; CSH1 141; and 'Oldambt' with 75 plants as check.

Meanwhile, the Compscan 3000 near infrared scanner (NIRS) had arrived and was calibrated with a series of steam distillation data. The non-destructive NIRS required c. 2 g samples, also allowing for replication, yet all seed was available for planting. The time necessary to prepare the samples after harvesting for scanning, the scanning itself and the selection procedure, however, made it impossible to produce a generation in one year.

Seed of the following numbers of plants were analysed in 1987: EOH1 146; EOL1 73; COH1 381; COL1 58; CSH1 96; Oldambt 50. Within a population selection was now one way, for example EOH was selected towards high %EO only. After ample consideration the selection for carvone content in the oil (CO) was changed into selection for carvone content in the seed (CS). The population CSH1 was to be developed towards an experimental variety (Exp.cv) and included plants of top EO and CS half sibs.

– The second generation was produced on plants sown in 1988 from the best 10–25% plants of selected half-sib families, keeping effective population size to a minimum of 30 parent plants. The fruit was harvested, cleaned and analysed in 1989.

– The third and final generation was similarly selected and raised in 1990 and its produce analysed in 1991. The results are summarised in Table 3 and it shows that the selection for high carvone content of seed (CSH) gave a consistently positive response.

– Field testing: The plant material described above had been raised in the protected environment of a cold frame and planted in field isolations as spaced plants, but they had not been grown as a crop. Performance in the crop phase, however, was considered to be the final test. Comparative field trials were carried out in 1992/3 on two sites. A treatment in the trials required several hundreds grams of seed. So as to meet this demand, proportionate amounts of rest seed in store of the families making up a population, were put together. This was possible with the second and third generations thanks to the use of the non-destructive NIRS method. The amounts of seed realised, however, were still not sufficient and it was decided to put the populations of EOH and CSH per generation together: EOH2+CSH2 was referred to as H (high) 89; EOH3+CSH3 as H91 and EOL3+CSL3 as L (low) 91, there was not enough seed of the second generation 'low' content (EOL2+CSL2) families.

Table 3 Average contents of populations RDMO

	%EO	%CO	%CS		%EO	%CO	%CS		%EO	%CO	%CS
Oldambt 1985	3.0	66	2.0								
	1st generation				2nd generation				3rd generation		
EOH1	4.2	52	2.1	EOH2	4.0	60	2.4*	EOH3	4.2	56	2.3
EOL1	4.3	53	2.2	EOL2	3.7	60	2.2	EOL3	3.7	63	2.3
CSH1	4.1	55	2.2	CSH2	4.6*	57	2.5*	CSH3	4.6*	60	2.7*
CSL1	3.6	50	1.8	CSL2	3.8	59	2.2	CSL3	4.1	57	2.3
Exp.cv1	4.8	48	2.3	Exp.cv2	3.7	62	2.2	Exp.cv3	4.5	54	2.4
Oldambt	4.0	54	2.1	Oldambt	3.9	55	2.1	Oldambt	4.1	54	2.2

* indicates significant difference with Oldambt; EO = essential oil in the seed; CO = carvone in the oil; CS = carvone in the seed.

The seed produced in the field trials was analysed by the Infralyser 500 NIRS machine calibrated with gas-chromatographic data whereas the selection was based on values generated by the Compscan NIRS calibrated with steam distillation data.

– Results: The field trials and the results were reported in detail by Toxopeus *et al.* (1996). Briefly summarised, both 'high' populations had a significantly higher carvone content of the seed than the 'low' and the parent Oldambt in both trials. The 'low' population was substantially and significantly lower than 'high' and marginally lower than Oldambt. Yields both in terms of weight of seed as well as carvone, of the 'high' populations were amongst the highest but not significantly higher than the average of standard varieties.

These results show and confirm that genetic improvement of carvone content of the seed by recurrent selection on a population basis can be effective without being accompanied by any loss of agronomic values.

7.3.3.2. Development of biennial experimental varieties

– The RDMO 'Exp.cv3' population (Table 3) was second best in contents although not significantly. More selection was apparently necessary. The 'Proefras 1' (experimental variety nr.1) was based on the seed of 93 plants selected for high contents from some 300 robust, high yielding plants selected from the former plot. The subsequent multiplication produced in 1996 showed carvone contents in the range of 2.6–2.7 compared to standard varieties in the same observation field of 2.1–2.3, an improvement of c. 20%.

– Applying the clonal propagation technique described in Chapter 7.2.3, 459 cuttings belonging to 44 clones were successfully established in the field in 1993. The next year 92 seed bearing plants from 22 clones were screened for contents, selecting the best 20 plants belonging to 10 clones. This seed was sown in 1995 for multiplication as 'Proefras 3'. Carvone content in 1996 seed was between 2.9 and 3.1, an improvement of nearly 50% on standard varieties.

– Recurrent population improvement for contents was also applied to accessions of NS material from eastern European countries, preparing for a possible successor to Bleija. Two selection cycles were carried out. This material, referred to as 'Proefras 4' does not, so far, show improved contents. More selection appears to be required, provided there is sufficient genetic variation.

7.3.3.3. Spring Caraway

– The variety 'Karzo' was bred from 4 accessions acquired in 1986, one each from Hungary and Egypt and two ones from Poland. Two years of observation and multiplication followed by two years of comparative field testing on different soils, and a winter of detailed observations in the glasshouse showed the differences between the 4 accessions to be marginal. They probably have a common origin in the Middle East. The bulked seed of the 1989 field trials was used for commercial testing in the field and factory (Toxopeus and Bouwmeester 1993). Variety registration was granted in 1993 under the name 'Karzo'. Spring caraway is acknowledged as a distinct crop type in the Rassenlijst since 1993.

Karzo can be sown in early spring, will survive a spring frost of −5°C, flowering takes place in July and the NS seed is ready for harvesting in late August–early September. Usually Karzo yields 1–2 t/ha of a sizeable seed (1000-s weight 4–5 g) and a %CS of 1.1–1.7% depending on soil type and year of production (Toxopeus and Lubberts 1994).

– The variety 'Springcar' was bred from the annual selection (SE) issuing from the F1 generation of the crosses between spring types and winter variety Bleija made in 1986 as explained in Chapter 7.2.1. In each of the following years the population was continued by harvesting early ripening and robust annual plants in the field but biennial plants kept segregating. In field plots of generations SE-r3 (3rd generation recurrent selection) and SE-r4 roughly half the number of plants remained vegetative and there was no sign of response to the selection for annuality. Apart from the satisfactory field performance, the selections produced a seed substantially higher in carvone content than that of Karzo. In the winter of 1990/1 a population of 179 spaced out plants of SE-r4 were grown in a greenhouse, in soil, at a temperature of 16°C, provided with assimilation light during 16 hours a day. Five months after sowing only 2 plants remained vegetative showing that, in the field, most vegetative plants are in fact annual but late types. The 43 earliest plants were selfed and all but two produced seeds. The full sibs were sown the next spring in an isolated field, together with rows of spring caraway for comparison as well some 100 SE-r4 half-sib families. The progenies with none or few vegetative plants were allowed to flower and fruit, late flowering plants being removed. The plots with spring type plants were removed prior to anthesis. The resulting seed was multiplied and field tested comparing to the SE-r4 and Karzo in two field trials one on sandy soil, the other on sandy loam. The selection and SE-r4 had a significantly, c. 15% higher carvone content than Karzo, this, combined with a higher seed yield resulted in a significantly, c. 30% higher carvone yield. (Toxopeus et al. 1996).

The selection has been submitted for registration as the variety 'Springcar' in 1995. The great advantage of annual caraway is that seed may be produced within a year of a new demand. Moreover the crop has proven to be a very suitable companion crop of the winter type.

7.3.3.4. Autumn-sown Winter Caraway

The very late annual plants observed to segregate in the SE-r4 as described above gave the impression that they might have flowered more quickly and fully with some additional cold induction. Such plants were thought to require less 'cold' for vernalization, and could possibly be cold-induced at a younger stage and with smaller taproots than plants of traditional biennial varieties. Whether these plants would be able to survive a winter in the field was the question. The idea was put to the test by conducting a small field selection experiment. Biennial Bleija, annual Karzo and SE-r4 were sown in equally large subplots, in early August 1992. In the event all Bleija plants did remain vegetative all through, Karzo was killed outright in the winter. The plant stand of SE-r4 was decimated during winter but the following spring showed about 50 flowering plants, from early to late, and some vegetative plants. Seeds that were about ripe by early August 1993 were harvested. This seed was dried and sown a fortnight later.

Once again many of the resulting plants died in the ensuing winter but a much larger proportion flowered in the following spring. About 50 of the earliest and best seeding plants were harvested. This seed was multiplied and put to commercial field test in 1996 which unexpectedly failed because the seed did not germinate. This dormancy problem is probably a result of the very short period of 5–6 weeks between harvesting the seed and sowing the next crop, and needs a solution before this 'autumn-sown winter caraway' is ready for developing varieties and cultivation.

The first indications are that yield and seed weight are at the same level as those of the traditional winter type. Carvone content is more like that of spring caraway, i.e. between 1.3% and 1.6%. Apart from maximising adaptation, improvement of the last trait would likely be the object of continued genetic improvement.

The added value of this new autumn-sown caraway crop type lies in its special ability to be sown in the autumn which allows it to be easily incorporated in prevailing field cropping systems in Europe. It is much better adapted to agro-economic conditions than the traditional biennial varieties.

7.3.3.5. The Tetraploid Population SVP 4X

The Bleija tetraploid population (see also Chapter 7.3.2) was subjected to population improvement for improved fertility for two generations. Out of c. 300 spaced plants in a field isolation the 10% plants with the best seed set were selected. This procedure was repeated in another generation after which the seed of the selected plants coded 'SVP 4X' was multiplied for field testing.

SVP 4X produces a dense, rather short crop with prolific fruit setting. However, at the time of filling the sterility phenomenon begins to show in that the embryo fails to develop in about half the fruits which develop parthenocarp, 'empty' fruits consisting mainly of husks, the exocarp. The empty fruits are blown away in the processes of cleaning and threshing. The weight loss is partially offset by the great weight of the filled fruits. Crop yields are between 700 and 1000 kg/ha. However, as the following analysis shows the husks do contain oil!

A sample of one kg of harvested produce was collected prior to cleaning. Fruits of the caraway plant occur in pairs (Figure 1b) that split when drying after maturity. Empty fruits occur either in pairs or together with a filled fruit in which case the two will not split. For the purpose of the study such fruit were separated by rubbing hard. It was noted that this was accompanied by a strong smell of caraway essential oil emanating from vittae burst in the rubbing process. The loss of oil was biggest in the filled seeds which were rubbed hardest as also became clear subsequently when oil content was measured. The sample thus prepared was passed over a number of sieves. The light, empty fruit fraction was further divided by air currents at different speeds. A sample of the original lot showed a 1000-s weight of 4.3 g and an oil content of 5.4%, a high figure considering steam distillation. It was estimated to contain about 200.000 seeds, a number that is also encountered in diploid fully fertile varieties, but only roughly half the number were filled.

The filled fruit fraction was separated into two with respectively 7.50 and 5.80 g 1000-s weight and with an estimated average oil content of 5.0%. The empty fruit fraction was split into three groups, the largest of which with a 1000-s weight of 1.4 g containing 9% oil and the others respectively 0.93 g and 6%, and 0.74 g and 2.7% Therefore, for purposes of oil extraction, the husk fraction should be included. At a yield level of 1 t/ha of the not-cleaned-not-threshed harvest, the oil yield would have been in the order of 54 kg/ha, almost competitive with the production of a regular variety.

It must be realised that the cleaned filled fruit is probably too big and spicy for the human palate, and therefore is unlikely to fit the existing 'whole seed' market, but there may possibly be unexpected alternative uses.

7.3.3.6. Germplasm collected in The Netherlands and SE Germany

CARAWAY IS BACK AGAIN was a headline (translated) in a local newspaper (Anonymous 1991) announcing the rapid expansion of a natural caraway population first discovered some five years ago in a nature reserve along the river Rhine near the village Neerijnen.

Until the 1950s caraway occurred in grassland vegetations almost everywhere in The Netherlands, also due to the fact that a little caraway seed was included in grassland mixtures. It was considered a healthy addition to the grass mix, as it still is in southern Germany. In The Netherlands caraway has become extinct in meadows because of the overall intensification of grassland management. But this is not the case in Germany where caraway is a highly regarded species ('Wiesenkümmel') common in zero input meadows.

Grown as spaced plants in a field isolation, the population found in Neerijnen was confirmed to be a biennial population. Plants were growing out sideways more than upwards, blooming and shedding their seed earlier and over a much longer period of time than their cultivated brethren in similar conditions. In a comparative yield trial 'Neerijnen' consisted of rather loose and short plants, early to start shedding seed, poorly yielding, and its seed with an average carvone content.

In July 1994 natural caraway was systematically collected in Bavaria and Frankenland in Germany yielding 12 populations of 'Wiesenkümmel'. A population consisted of up to six samples collected in an area around a town where caraway was observed, in meadows, but also as a weed on a parking lot or on dirt roads. A sample was collected by walking a wide circle through the place where caraway was spotted, picking at least 30 plants with the largest fruits. Plants with ripe shattering seeds were hardly encountered because the meadows would be cut 6 or 7 times in the growth season for its highly regarded aromatic hay. So the seed collected was nearly mature at best.

Carvone content of the seed is stated in Table 4. The contents are high and some were almost unbelievably high (6.2% in a sample from Freising). However, after having grown the samples in Wageningen, the contents converged to a range of 2.4–2.9%, the same range as the values of the standard varieties grown in the same field. Nevertheless, the Freising population still is amongst the highest with a sample at 3.4, and an average of 3.1; and the one from Weyarn is steady showing a range of 3.1–3.4 both as original sample and as progeny. Strikingly, Zijlstra (1916) reported a similar phenomenon (see also 3.1) 80 years ago as shown in Table 1. The phenomenon is discussed in the next subchapter.

Table 4   Carvone content of accessions collected in SE Germany, and
their progenies grown in The Netherlands

Collection	Original samples			Progenies grown in NL.		
	nr. of samples	range	mean	nr. of samples	range	mean
Wernsbach	6	2.4–3.3	2.6	5	2.4–2.9	2.7
Pappenheim	6	2.7–4.0	3.8	6	2.4–2.8	2.6
Holzkirchen	4	2.4–4.4	3.2	4	2.8–3.3	3.0
Jachen	4	2.6–3.6	3.1	4	2.2–3.0	2.6
Andechs	2	3.4–3.7	3.6	1	3.0	
Freising	4	2.7–6.2	4.4	2	2.7–3.4	3.1
Weyarn	2	3.1–3.4	3.3	2	3.0–3.4	3.2
Spitzingerse	2	3.3–3.4	3.4	2	2.9–3.0	3.0
Schliersee	4	2.7–4.2	3.2	3	2.8–3.0	2.9
Gotteszell	3	3.0–3.2	3.1	3	2.5–2.6	2.6
Wiesing	2	2.7–3.4	3.1	2	2.1–2.5	2.3
Kirchhoff	3	2.7–3.7	3.1	2	2.5–2.7	2.6
Standard vars				5	2.4–2.9	2.8

7.4. YIELD AND SEED VALUES AS WELL AS INTERACTIONS

Yield, unless otherwise specified, is the weight of threshed seed with a moisture content of c. 10% per ha.

Seed (fruit) values are (1) *weight*, expressed as the weight of 1000 seeds, (2) *contents* of essential oil and/or carvone expressed as percentage of total seed weight (moisture content to be specified), and (3) the *volume of oil* in the seed here expressed as '1000-s volume' in μl. All three are elements of the same seed, the fruit, and it is important to understand the underlying relations. Finally, there is *quality* which definition, however, depends on the user of the product. The consumer of the seed in bread products wants a reasonably pungent well filled seed for a good bite, the standard of which is the seed of the traditional winter type. Quality for the distiller is a matter of cost and benefit regardless of size and appearance of the seed provided the product is typically caraway essential oil.

The data presented in Table 2 provide a unique picture of performance of the biennial caraway archetype, the variety 'Noord-Hollandsche' grown in the heavy clay of the Oldambt area of north-eastern Groningen. Apart from providing information on annual effects over a period of nine years, the data on seed values provide a unique research opportunity to study their interrelations.

7.4.1. Annual Effects

Yields fluctuate between 1 and 2 t/ha with the 465 kg/ha of 1933 being exceptionally low. This is probably the result of a very 'hollow' stand with many vegetative plants that were too small in winter to be effectively vernalised, one of the hazards of growing biennial caraway. Farmers believe that a hollow stand produces a seed with a high oil content, which seems to be confirmed in this case with both oil and carvone content amongst the highest of the series. Thousand seed weight varies between 3 and 4.4 g, with that of 1934 being exceptionally high and with an equally exceptionally low oil content as if carbohydrate supply had been directed disproportionally towards embryonic tissues.

Oil content fluctuates around the 4% containing an average carvone content of about 53%. By far the highest oil yield of 103 kg/ha was observed in 1936. This was the year of the highest seed yield combining with an good average 1000-s weight and an average oil content, but with the lowest % carvone in the oil.

1000-s volume was highest in 1933: 205 µl giving an indication of the capacity of this sink. All the same, however, in 1934, the following year, it was 51 µl, only 25% of the former capacity! Are the vittae so much smaller or do they remain partly empty?

Seed and carvone yields/ha reported some 60 years later by Toxopeus and Bouwmeester (1993) and Toxopeus and Lubberts (1995), using varieties like Volhouden, are completely in line with the above data except that they show much less variation in yield, undoubtedly the result of intensive field trial management without the use of a companion crop.

7.4.2. Effect of the Date of Cutting the Crop

The most striking effect of the date of cutting is on the oil content: the later the date of cutting the plants in the field the lower the contents in the seeds. This effect occurs in all years recorded and is seen most clearly in the data of the last three years of Table 2 (column 4). The other value systematically affected is 1000-seed weight (column 3) increasing with advancing cutting date very probably because the embryo inside the fruit is filling rapidly in this period.

Volume of oil, shown in the last column, tends to be stable within a year, it is not affected by the date of cutting the plant, as is best illustrated by the year 1937. Considering that 1000-s volume is the outcome of multiplying *1000-s weight* with *oil content* (and specific gravity), this means that the decline of oil content in time was proportional to the increase of seed weight in the same period! Oil filling, within the limits of that year, had apparently been completed before the first date of cutting.

The observation that carvone content of the oil (column 5) tends to be stable within the year, shows that the process of oil formation, too, was completed before the first cutting date. The conclusion is that degree of maturity of the fruit is a determinant of contents. *The more mature the seed the lower its contents.* This phenomenon probably explains the high contents recorded in the seed of various degrees of immaturity collected in S. Germany (Table 4). Seed from plants subsequently grown in Wageningen was ripe when analysed by NIRS, explaining both the lower values and also the smaller variation in contents.

7.4.3. A Note on 1000-s Volume and the Volume of the Vittae

The organ containing the essential oil in the fruit are the vittae, of which there are 6 in the exocarp of each seed. Dividing 1000-seed volume by 6000 gives the average volume of one vitta.

Winter caraway varieties with an average 1000-s weight of 3.5 g, and oil content of 4% (by NIRS) have a 1000-s volume of 156 μl.

Spring caraway Karzo with a 1000-s weight of 4.5 g and an oil content of 3% has an average 1000-s volume of 150 μl. The volumes of the two types are practically the same.

However, the volume of tetraploid seed is a different story. 1000-s volume of filled seed at an average 1000-s weight of 6 g and an oil content 6% (by NIRS) is 400 ml, nearly three times the former, diploid, values. The 1000-s volume of the empty seeds is much smaller. At an average 1000-s weight of 1 g and a content of 8% its volume is only 90 ml, less than 25% of the volume of the filled seed. Vittae development is apparently stimulated by the growth of a viable embryo in the fruit.

7.5. CONCLUSIONS

The results and success of the recurrent population improvement RDMO have made it abundantly clear that caraway is typically a cross-fertilising species. Populations improved for seed values or quality traits may be increased as varieties, and will remain a solid breeding foundation for subsequent action, be it for the development of clones or lines which, after selection may be recombined to form the basis of new varieties.

Phenomena such as inbreeding depression and heterosis are to be expected once selfing is employed to develop lines. It is interesting to note that botanical mechanisms regulating mating system, such as male sterility or self-incompatibility have not so far reliably been reported.

The near endless series of data on content generated by the NIRS machines were always screened for evidence of possible genetic variation for an increased proportion of carvone in the oil. This has never been encountered, although there were several occasions of excitement. It was the background for the decision to try to develop a protocol for genetic transformation that was recently completed (Krens et al. 1997), providing for the opportunity to introduce the necessary gene(s) from other species. Ultimately, this transformation, combined with several generations of seed breeding and clonal selection, could develop a synthetic variety containing say 80% of carvone in the oil rather than the 55% in present varieties.

The yield objective would be to recombine the existing top level of c. 6% oil content of the best populations with the recombinant material with 80% carvone in the oil, without loosing agronomic qualities. The result would produce a variety with a seed containing 4.8% carvone, capable of a carvone yield of nearly 100 kg/ha at a yield of 2 t/ha.

The case of the SVP4X has shown that autotetraploidy can produce remarkable products. However, sterility will not simply disappear with increasing number of generations, despite selection for good seed yield of individual plants. Sterility in SVP 4X will have to be thoroughly studied before effective action may be expected. Moreover market studies should survey the prospects of the large seed produced.

In case sterility could be controlled, the prospects for commercial oil production is a population that would produce 90% or more empty seeds with an oil content of 8–10%, i.e. 4–5% carvone. An average yield of 1 t/ha would thus produce 50 kg carvone/ha which is more than the average yield of 1.5 t/ha of traditional winter caraway with a carvone content of 2% producing 30 kg carvone/ha, and in addition produce 50% less of waste!

Developing spring caraway as a crop has doubled opportunities for caraway production in Europe. Cultivating spring caraway makes fresh seed available to the market within about 8 months upon making new demand worthwhile. For the traditional winter type, this period is about 18 months, which is one of the causes for extremely high prices in situations of shortage, inviting subsequent overproduction and years of chronically depressed prices. Because spring caraway is a good companion crop of winter caraway probably makes it practical to cultivate caraway in a two year period in crop rotations.

The new autumn-sown winter type would provide seed within 12 months in response to the market. If varieties of this type could be developed that would be equal in seed quality and yield to the traditional winter type, it might well take the place of both former crops. First because it would yield more than the combination of the other two, and secondly because it would fit very well in existing rotations. It is agro-economically the most cost effective.

Efforts to genetically improve caraway for traditional seed production should concentrate on the improvement of productivity since the quality of its product, the seed, is traditionally fixed and therefore cannot be improved.

The main constraint of the traditional winter type is the need for growing it under a companion crop. Breeding varieties with a higher degree of resilience to the rather harsh conditions of juvenile plant development under the companion crop may be a selection priority.

However, efforts are probably best invested in developing durable varieties of the autumn-sown winter type because of its basic superior agronomic adaptation to field cropping in NW Europe.

ACKNOWLEDGEMENT

The support and assistance of Dr. J. Hoogendoorn of CPRO-DLO in developing and finalizing this document is gratefully acknowledged.

REFERENCES

Allard, R.W. (1960) Random mating in small populations. In *Principles of Plant Breeding*, John Wiley and Sons, Inc., pp. 200–203.

Anonymous (1991) *De Gelderlander* of 13.7.1991.

Bouwmeester, H.J. (1995) Seed yield in caraway (*Carum carvi* L.) 1. Role of pollination. *J. Agric. Science*, **124**, 235–244.

Broekema, C. (1924) Aanbevelenswaardige gekweekte rassen van landbouwgewassen (Recommended bred varieties of field crops). *De Veldbode*. 13.9.1924, 359–360.

Dijkstra, H. (1993) Personal communication.

Dijkstra, H. and Speckmann, G.J. (1980) Autotetraploidy in caraway (*Carum carvi* L.) for the increase of the aetheric oil content of the seed. *Euphytica*, **29**, 89–96.

Goedewaagen, M.A.J. and Zijlstra, K. (1925) Gewone, loszadige karwij en een nieuwe zaadhoudende varieteit. Een vergelijkend-anatomisch onderzoek der vruchten. (Common, shattering caraway and a new non-shattering variety. A comparative anatomical study of the fruits.) *Verslagen Landbouwkundig Onderzoek, Rijkslandbouwproefstations*, 30, 287–306.

Hegi, G. (1926) *Flora von Mitte... a* V-2, Munich, Germany, pp. 1181–1187.

Keulen, D. (1988) Kruisingsaspecten en de overerving van eenjarigheid bij karwij (*Carum carvi L.*) (Aspects of crossing and the inheritance of annuality in caraway). *In thesis* Plant Breeding Dept., Agric. University Wageningen, 44 pp.

Kops, J. (1840) *Flora Batavae*, The Hague, The Netherlands, Vol. IV, 267.

Krechting, C.F. (1970) Bepaling van het gehalte van etherische olie en carvon van karwijzaad. (Determination of contents of essential oil and carvone of caraway seed), 3 p., Foundation for Agricultural Plant breeding, SVP, Wageningen, unpublished report.

Krens, F.A., Keizer, L.C.P. and Capel, I.E.M. (1997) Transgenic caraway, *Carum carvi* L., a model species for metabolic engineering. *Plant Cell Reports*, 17, 39–43.

Oostrom, J.J. and Van der Kamp, H.J. (1990) Ontwikkeling van een gaschromatografische methode voor de bepaling van carvon en limoneen in karwij (Development of a gas-chromatographical method to determine carvone and limonene in caraway seed, with English summary). *Rapport* 90.58, RIKILT-DLO, Wageningen, The Netherlands.

Rassenlijst (1930) Karwij; (1936) Karwij; (1965) Karwij; (1972) Karwij; (1993) Karwij.

Toxopeus, H. and Bouwmeester, J.H. (1993) Improvement of caraway essential oil and carvone production in The Netherlands. *Industrial Crops and Products*, 1, 295–301.

Toxopeus, H. and Lubberts, J.H. (1994) Eerste ras zomerkarwij krijgt kwekersrecht (First spring caraway variety gets plant breeders rights). *Prophyta*, 1, 18–19.

Toxopeus, H., and Lubberts, J.H. (1995) Effect of genotype and environment on carvone yield and yield components of winter-caraway in The Netherlands. *Industial Crops and Products*, 3, 37–42.

Toxopeus, H., Lubberts, J.H., Neervoort, W., Folkers, W. and Huisjes, G. (1996) Breeding research and in vitro propagation to improve carvone production of caraway (*Carum carvi* L.). *Industrial Crops and Products*, 4, 33–38.

Tusveld, M.A.H., Frankhuizen, R. and Van der Kamp, H.J. (1991) Onderzoek naar de bepaling van het carvon- en limoneengehalte in karwijzaad met NIRS (Determination of carvone and limonene in caraway by NIRS, with English summary). *Rapport* 91.34, 18 pp., RIKILT-DLO, Wageningen, The Netherlands.

Van Roon, E. and Bleijenberg, H.J. (1964) Breeding caraway for non-shattering seed. *Euphytica*, 13, 281–293.

Wander, J.G.N. (1994) Teelt van karwij (Cultivation of caraway). *Teelthandleiding* nr. 60, 40 pp. PAGV, Lelystad, The Netherlands.

Wit, F. (1989) Personal communication.

Zijlstra, K. (1916) Ueber *Carum carvi* L.. *Recueil des Travaux botanique Neerlandais*, Vol. XIII, Livres II et IV., 159–340.

Zijlstra, K. (1940) Het verband tusschen zichttijd en opbrengst van karwij (The relation between time of cutting the caraway crop and its yield). *Vereeniging tot exploitatie van de proefboerderijen in de provincie Groningen, verslag over de jaren 1935–'39*, pp. 137–141.

8. PRODUCTION OF BIENNIAL CARAWAY FOR SEED AND ESSENTIAL OIL

ZENON WÊGLARZ

*Department of Medicinal Plants, Warsaw Agricultural University,
Nowoursynowska 166, 02-787 Warsaw, Poland*

8.1. GROWING AREAS

Caraway (*Carum carvi L.*) occurs as wild plant within phytocoenoses of meadow type, usually in a humid coastal or mountain climate (Nowiñski 1959). Such ecological conditions should be taken into account in the choice of region for caraway cultivation, and also to a certain degree, in the choice of soil type and a sequence in crop rotation. In Europe, where this species has been widely grown for over 200 years, caraway was and usually still is produced in coastal regions of the Netherlands, Germany and Poland as well as in a piedmont of the Czech Republic and Slovakia.

The second factor that must be considered in caraway cultivation is the character of its development, since it could be either biennial or facultatively perennial crop with a distinct juvenile phase (Chládek 1969, 1974, Jankulov 1960, Novák 1973, Wêglarz 1982). Besides, there are some caraway properties typical of a wild plant, that favour its adaptation for the development in natural sites, having however a negative effect on its cultivation. Elimination of such traits is difficult, particularly in relation to non-uniform seed germination, uneven fruit ripening on plants and shattering.

8.2. SOIL REQUIREMENTS, SEQUENCE IN CROP ROTATION AND FERTILIZATION

Caraway as a plant of high soil requirements grows and yields best on deep and warm soils, rich in humus and nutrients. The most suitable are fen soils, loess, chernozem, limestone soils and deep but not too heavy clays (Chotin and Szulgina 1963, Heeger 1956; Rumiñska 1990).

The best forecrops for caraway are considered root and vegetable plants previously supplied with a full rate of farmyard manure (20–40 t/ha). Suitable are also clover, lucerne and other mixed papilionaceous crops. Besides, plants ploughed-in for green manure could also be recommended.

In contrast, cereals are considered the least suitable forecrops for caraway. On the other hand caraway itself performs well as a forecrop for cereals, by leaving a field almost weed-free, and above all, due to its early harvesting there is enough time to conduct all pre-sowing agricultural practices, so essential for cereals (Chotin 1959, Rumiñska 1981).

Caraway is highly sensitive to soil water, considering the level of that underground and soil-bound. Too high water table and stagnant water in particular are dangerous especially in spring and may cause mass wilting of the plants. A relatively high, stable soil moisture is necessary for the adequate development of caraway since it originates from wet meadows.

According to Buszczak (1962) and Węglarz (1983/a), soil moisture between 50–80% is optimum for caraway plants, ensuring their harmonious growth and the highest yield of the fruits. Under such conditions the plants use evidently less water to produce a particular weight of the fruits and the best weight ratio of fruits to straw is observed. At extreme levels of soil moisture there is more non-flowering plants in the second year, while those with already formed generative organs show a delayed flowering and more abundant umbel malformations (Węglarz 1983/a). Upon the experience and long-time observations by caraway growers in Poland this crop should be considered particularly sensitive to water deficiency during germination and emergence, being comparable to spring cereals.

Caraway grows well only on neutral or slightly alkaline soils (Chotin and Szulgina 1963, Heeger 1956). Rumińska and Kaczor (1963) proved that an increased pH through adequate liming enhanced the yields of both fruits and essential oil. Calcium fertilizers are usually applied either before a forecrop or in late autumn preceding spring sowing.

Soil nutrients pertain to the factors highly essential for the process of caraway yielding, having both indirect and direct effect. The first is reflected in the influence of fertilizers on the size of plants at the end of the first year, that is a decisive factor for their potential vernalization and in consequence, flowering and yielding in the following season. Initially, Kramer (1955) pointed that the increased rates of nitrogen in the first year of caraway cultivation brought about more flowering plants and a higher fruit yield next season. Such a relation was also evidently shown by Węglarz (1983/a). In a pot experiment the plants poorly nourished in the first year were too small to form generative organs next spring (Tables 1, 2). Instead, an increased NPK fertilization in the first year resulted in significantly larger plants, being subsequently able to flower and yield. The same author (Węglarz 1982) tried to determine the relationship between the size of caraway roots and its productivity. Rootstocks classified into 3 grades: thick (root diameter 16–25 mm, fresh weight 51–150 g), medium (diameter 8–15 mm, fresh weight 26–50 g) and thin (diameter 2–7 mm, fresh weight 4–25 g) produced the plants which flowered in 100, 85 and 65% respectively (Figures 1, 2). Another interesting relationship appears

Table 1 Effects of fertilization and soil moisture
on the percentage of plants with floral sprouts

A level of mineral fertilization	Soil water capillary capacity (%)			
	30	50	70	90
$(NPK)_0$	0.0	3.0	0.6	0.0
$(NPK)_1$	34.0	41.0	25.0	19.0
$(NPK)_2$	53.0	76.4	72.0	28.0

$LSD_{0.01} = 3.91$.

Table 2 Effects of fertilization and soil
moisture on the yield of fruits (g/plant)

A level of mineral fertilization	Soil water capillary capacity (%)			
	30	50	70	90
$(NPK)_0$	0.00	0.00	0.00	0.00
$(NPK)_1$	0.98	1.24	0.84	0.50
$(NPK)_2$	3.02	4.84	4.36	0.90

$LSD_{0.01} = 0.88$.

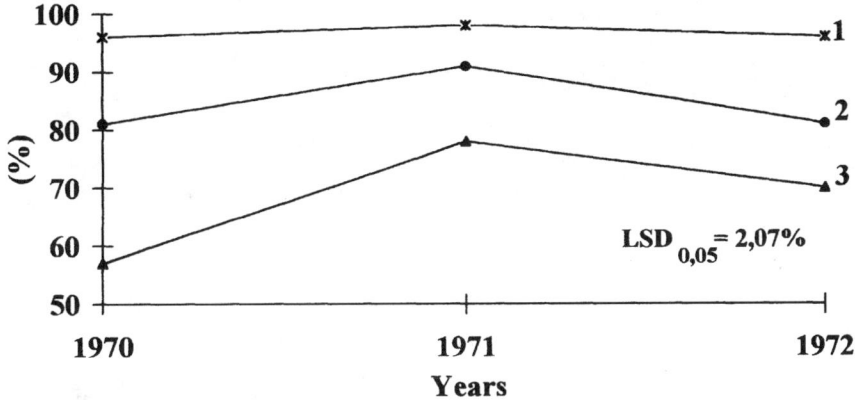

Figure 1 Effect of root size on the number of floral sprouts (%): 1 – Thick roots; 2 – Medium roots; 3 – Thin roots

from Figure 1. During the 3 year research all the plants obtained from thick roots formed umbel shoots. At the same time flowering of the plants obtained from medium and thin roots in particular differed within the years, proving a considerable influence of climatic factors on development of caraway. Seed plants obtained from thick roots were bearing substantially more fully-formed seedstalks and showed a higher weight of both total and 1000 fruits. However, there were neither significant differences in germination capacity nor in essential oil content in the fruits.

Direct influence of soil nutrients on caraway yielding occurs in the second year of cultivation, however, it has never been subject of a detailed investigation. Both in researches and agricultural practice the effect of fertilization is regarded as whole for two growing seasons.

Nutrient intake by caraway plants is intensive estimated by Schröder (1964) as 85 kg N, 39 kg P_2O_5 and 94 kg K_2O per ha, to yield 1.2 and 4.2 tons of fruits and straw respectively.

There is a positive response to both organic and mineral fertilization. The most intensive intake of nutrients is observed during summer and autumn in the first year and also in early spring of the following season (Rumiñska 1981). Detailed study

Figure 2   Roots of caraway: 1,2 – Large; 3,4 – Medium; 5,6 – Small (See Color Plate I)

conducted by Kordana *et al.* (1983) focused on the dynamics of nutrient intake and revealed that most nitrogen was absorbed by caraway plants at the rosette stage, i. e. July–September of the first year and during the formation of umbel shoots in the following season. Potassium intake was high in both years of cultivation, although it particularly enhanced in the second season while the plants were forming umbel shoots and during flowering. Maximum intake of calcium and phosphorus took place in September and October of the first year, then during fruit formation and ripening.

Most authors researching into caraway fertilization consider nitrogen the most essential nutrient that determines the level of fruit yield (Boshart 1942, Czabajski *et al.* 1960, Mihalea 1965, Nordestgaard 1986, Scharrer and Lisner 1965, Schröder 1964). Fertilizer formulation and its quantity considerably depend, as in most crops, on the type and soil fertility as well as on forecrop, weather conditions etc.

Experiments conducted during 1950s in the former Soviet Union (Ziemlinskij 1954) showed a very strong positive relationship between a direct fertilization with farmyard manure and caraway fruit yield. Moreover Aflatuni *et al.* (1993) found that composted manure increased essential oil content in the fruits. Caraway crops in Poland are often provided with farmyard manure (20–40 t/ha) which is usually applied prior to forecrop. However, the main source of nutrients are mineral fertilizers, supplied in both years of cultivation. In northern and central Poland, where the plantation of this plant is most concentrated and cultivated on fen and chernozem soils, pure crop is usually provided with the following rates of fertilizers: 60–80 kg N, 70–80 kg P_2O_5, 100–120 kg K_2O and 20–30 kg MgO, applied both in the first and second growing season. In the case of mixed cultivation, in spring of the first year a cover crop is supplied with its full rate of fertilizers. At that time caraway is provided only with half of recommended amount of nutrients while the other half is given after a cover plant is harvested. Fertilization applied in the second year follows the recommendations for a pure crop (Rumińska 1990). The above rates of fertilizers correspond with those suggested by Nordestgaard (1986).

Timing of mineral fertilization seems to be particularly important. Full rates of PK compounds together with half amount of N-fertilizers are applied prior to sowing in late autumn or early spring. The other half of nitrogen is provided after caraway emergence. Some growers supply nitrogen compounds in three rates, especially in a mixed crop, giving the final portion after cover plant harvest. In the second year of cultivation the entire PK fertilization and half of N-compounds are applied in early spring as soon as possible. The remaining part of nitrogen should be supplemented 2–3 weeks later.

N-fertilization, as mentioned before, is highly effective but, if incorrectly performed, it may cause considerable damage, especially if applied at excessive rates or too late. Such negative effects in the first year may bring an exuberant vegetative growth of caraway plants, that subsequently disturbs their harmonious entry into winter dormancy. In the second season, consequences can be more serious. At a long-lasting high soil moisture, due to heavy precipitation, caraway flowering is prolonged, that subsequently causes non-uniform ripening of umbels and fruit shattering. Excessive N-rates can also lead to substantial losses if during fruit ripening there are high temperatures and strong insolation, and in consequence – physiological drought. Under such circumstances, even immature green plants wilt in a very short time. A considerable part of fruits may not be properly formed, thus reducing fruit yield, seed germination and essential oil content. The latter was proved in a long-time observations by the author of the present paper. It seems particularly important, considering the observations of recent years which show a constant decline in essential oil content in fruits from caraway plantation.

At present in most European countries, multicomponent fertilizers of adequate composition for particular crops are in common use. For caraway, mixed formulations with a substantial content of magnesium are very suitable, particularly on light soils. Besides, Polish growers observed a positive effect of boron on both seed yield and essential oil content in the fruits.

Kordana and Zalęcki (1993) examined a multicompound liquid formulation (Agrosol U) used for foliar feeding of caraway plants. At full mineral soil fertilization, there was an additional positive effect of such treatment at the rate of 6–8 kg/ha, expressed by 20% rise yield, including an increase in both 1000 fruit weight and essential oil yield. These results show a high biological potential of caraway plants and should prompt further research to enhance productivity of this crop.

8.3. SOIL PREPARATION, SETTING UP OF A PLANTATION

A field for caraway should be prepared with special care, considering deep penetration of its roots and also a period of 2 or even 3 years of its cultivation at the same place. Ploughing should be **necessarily** performed in autumn, since long-time observations provided evidently worse results if it was conducted in spring (Rumińska 1990). On deep soils there is recommendation to use standard plough provided with a chisel, or a special subsoil cultivator, working askew or across the direction of basic ploughing.

In spring, agricultural practices should be carried out as soon as possible to prevent clodding and drying out of the soil top layer, and they include dragging and multiple harrowing. Cultivator is recommended only for very heavy soils, strongly stale after winter. This operation should be immediately followed by harrowing. If the soil is too loose and uneven on the surface, the use of a flat roller is required for precise sowing and to enhance capillary rise of soil moisture, and finally to obtain uniform emergence of caraway seedlings.

Caraway sowing material are achenes, whose germination capacity should be at least 80%. This level can be maintained for 2–3 years, then rapidly declines (Bocheńska and Kozlowski 1969, Heeger 1956). The weight of 1000 achenes may reach 5 g and according to commercial standards in Poland, it should not be below 2.6 g. In spite of a high germination capacity in a laboratory, caraway seeds sown in the field may germinate unevenly, taking a long time. The research carried out during the 1950s and 1960s showed that the field germination capacity could be increased by a pre-sowing stimulating treatment. Vrzalova (1955) achieved a better field emergence by soaking achenes in water (running in particular) and explained such an effect by washing out seed germination inhibitors. Chotin and Szulgina (1963) obtained significantly advanced and more uniform emergence as a result of seed stratification at 0°C for 20–25 days and also by warming up the seeds just before sowing. A sufficient water supply during germination is required for uniform emergence. Optimum soil moisture for caraway germination is 80% (Buszczak 1962).

At present, caraway is grown only from direct sowing. In the moderate climate of Central Europe two cultivation methods are used: mixed or pure crop. Difference concern only the first growing season. Mixed cultivation with a cover crop is usually preferred by the owners of relatively smaller farms located under favourable soil and climatic conditions. Those growers are orientated towards intensive usage of arable land. Cover crops are cultivated together with caraway and yield in the first year, being mainly represented by garden poppy, pea, black cumin and spring oilseed rape. Moreover, coriander is also considered by Polish growers as a cover crop very suitable for caraway. In Hungary the most popular cover crops are dill and annual caraway (Bernáth 1993). However, Müller (1990) is of a different opinion, regarding coriander as well as dill, marigold and spring barley grown for grain as the plants which have a negative effect on caraway performance. According to the same author, spring barely grown for green forage can be recommended as a suitable cover crop, likewise camomile, if sown together with caraway in previous autumn. Experiments carried out in Poland during the 1970s (Jaruzelski 1974, Kubis 1973) pointed also to flax and spinach as potential cover crops for caraway.

According to some authors (Novák 1973, Rumińska 1981) pure sowing can be delayed to the end of May, and even to August. However according to Węglarz (1983/b) and over the years of practice the growers recorded the highest yields from crops sown as soon as possible (Figure 3). In mixed cultivation, possibly the earliest sowing (March, April) is just obligatory.

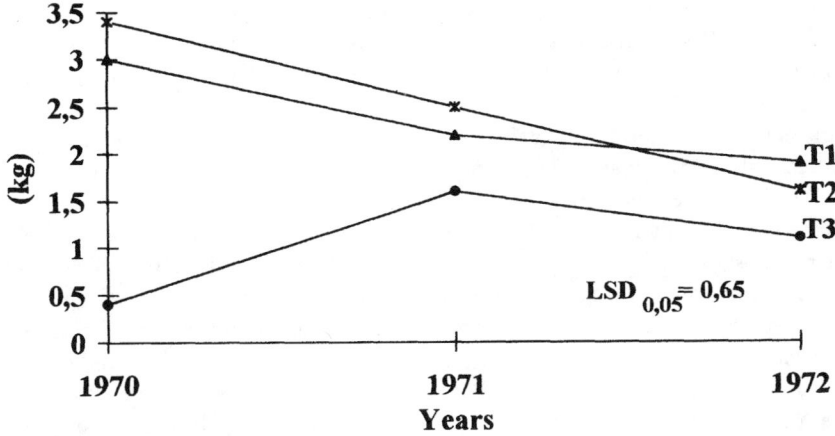

Figure 3 Effect of time of sowing on the yield of fruits (kg/plot): T1 – Early spring sowing (20–30. III.); T2 – Late spring sowing (20–30. V.); T3 – Summer sowing (5–15. VII.)

The sowing rate depends on both the cultivation method and soil type, ranging from 6–8 kg to 8–10 kg per ha for mixed and pure crops, respectively. A lower sowing density is recommended for firm soils, maintained in a good culture. According to Nordesgaard (1986), the amount of 4 kg of seeds per ha is sufficient for pure sowing in a single-harvest biennial cultivation, while the required amount for a double-harvest three-year crop rises to 6–7 kg. The sowing is performed in rows at a 35–50 cm spacing (larger on heavy soils rich in humus). Sowing depth increase from 1.5 cm on heavy soils to 2–4 cm on relatively lighter soils, exposed to fast drying of the top layer.

In a mixed crop the sowing rate is 20–40% lower as compared to pure cultivation. Long-time practice proved that cover plants (excluding pea) should be sown together with caraway into the same rows. Pea, instead, is sown into rows at a distance of 8 cm, followed by the sowing of caraway at a row spacing of 40–50 cm. In Hungary dill is sown in cross direction to the rows of the main crop, and before it (Bernáth 1993).

8.4. MAINTENANCE AND CARE OF PLANTATION

Appropriate timing and adequate quality of agricultural practices are particularly important for caraway, considering its as a biennial crop and also, as mentioned before, cultivation often combined with other crops. Such measures include loosening (aeration) of soil, weed control, thinning of cover plants and disposal of their post-harvest residues and, if necessary, thinning of caraway and ridging.

Seedlings emerge usually after 2 weeks and, like in other umbelliferous plants, their initial growth is slow. At that time the young plants are particularly sensitive to shading, and also to insufficient aeration and moisture of the soil. Such negative effects can be reduced through soil loosening as well as by chemical and mechanical weed control. The latter measure clearly differs from the loosening of the soil by the depth of penetration. Whereas weeding is shallow, to prevent the lifting of weed seeds from lower soil layers, the loosening is deeper, even down to 10 cm. Frequency of such practices depends on the soil and weather conditions, and differs for the first and second growing season.

Herbicides are commonly used in caraway cultivation, as it is relatively resistant to these chemicals (Pruszyński 1995, Müller et al. 1989). Since new products are being constantly introduced, selection of herbicides suitable for caraway also changes.

A plantation should be free of weeds during the first year in particular. In a pure crop herbicides can be then applied three times: first directly after sowing (linuron, prometryne, metobromuron), next at the stage of 2–3 leaves (linuron, metolachlor) and the final treatment for the first year in late autumn, before the soil gets frozen (propyzamide). In the second season, herbicides may be applied at the beginning of caraway vegetation. In both years monocotyledon weeds can be eliminated with the use of graminicides, regardless of the developmental phase of caraway plants (fluazifop-P-butyl). Chemical weed control in mixed cultivation with coriander follows the rules of herbicidal treatments in a pure crop. When caraway is grown together with flax or pea, Afalon provides good results (Kubis 1973, Pruszyński 1995). On the case of other cover plants, chemical weed control may begin in autumn of the first year and should follow the principles related to a pure crop.

While caraway is grown together with poppy, the thinning of this cover plant is very essential and should be performed as soon as possible. In late summer and autumn of the first growing season there is the most intensive gain in plant matter, therefore cultivation practices in that period are of a major importance. In mixed crops, post-harvest residues of cover plants should be instantly removed and weeds eliminated. In the case of high stubble (e.g. seed-coriander) such residues can be mechanically finely chopped and left in the field.

Another important measure is harrowing, carefully performed across or askew to caraway rows. Apart of loosening the soil, harrowing is a method for thinning the plants, especially if they are too dense. This procedure frequently combined with the use of N-fertilizers applied as top dressing, followed by mechanical loosening of inter-rows. Towards the end of the first year, shortly before the soil gets frozen, shallow ridging is often performed to prevent caraway plants from freezing-up during severe snowless winters, and also from damage caused by hares.

In spring of the second growing season, cultivation practices should be conducted as soon as possible. They mainly include loosening of inter-rows and weed control. Also, harrowing across caraway rows is often carried out at the same time. All these procedures are usually combined with mineral fertilization. If the plantation is intended for the third year, all post-harvest practices are analogue to those recommended for the first growing season, including mineral fertilization.

8.5. HARVESTING, DRYING, STORAGE

In Europe, caraway is harvested in the period from late June to mid-July. Appropriate time is particularly important, considering non-uniform maturation of umbels and the fruits highly susceptible to shattering. It appears from long-time observations that caraway harvest begin when in most plants the fruits of the main and top lateral umbels change colour from green to brown. A delay brings inevitable losses due to shattering of the most valuable fruits. On the other hand, if the harvest is advanced a lot of fruits will not ripen, being of a lower quality for sowing and row material.

A two-stage harvest brings the best results. Cutting with the use of a mower, reaper or a binder should be performed early morning (with dew) or on a cloudy day. Cut plants should be left in swaths or sheaves for a few (up to 10) days, before they are threshed. This short period from cutting to threshing is very essential, since then the fruits become finally formed and coloured. Warm weather favours this process, however too intensive insolation is unwanted. Observations by the author of the present paper revealed a rather peculiar, positive relation between rainfall over that period and the quality of raw material, expressed by a more uniform, light-brown colour of the fruits and their distinctly smaller contamination with caraway carpophores and with weed seeds. A combine harvester operating in the field is the best method for caraway threshing. Often, such harvesters are specially adapted for this crop (Hecht *et al.* 1992). Threshers are also used, however, transport of dry plants from the field usually increases yield losses. After threshing and mechanical cleaning, the fruits should be redried down to 10–12% moisture content. Then, for some time the fruits should be kept loose in a thin layer, being frequently mixed, within a dry and aerated storeroom to finally establish their moisture content. Such prepared raw material is packed into sacks and, if inadequately stored, it can go musty and mouldy, thus becoming useless as raw material.

Caraway yields widely fluctuate, reaching 1–3 t/ha. In mixed cultivation with cover crops, likewise on a plantation left for the third year, the yields obtained may be 15–30% lower (Müller 1990, Rumińska 1981).

8.6. PESTS, DISEASES AND THEIR CONTROL

Caraway pertains to aromatic plants grown for years in relatively large areas and its field production is often concentrated within specific particular regions. Under such circumstances this crop is more frequently attacked by pests and diseases than other medicinal plants. Pests are the major threat to caraway, especially if not controlled, they can cause a total loss of fruit yield.

Deppresaria nervosa Hav. appears to be the most serious pest. Adults overwinter under tree bark, in straw left in the field, in sheds etc. In the second year of caraway vegetation the females deposit eggs into leaf folds. An emerged caterpillar, which first is of light then dark-grey colour with a visible row of black mamillae with a white border, finally turns black. Initially, caterpillars feed on the leaves, then as the plant grows, they get into umbels covering them with web, destroying the flowers and newly formed fruits. When their feeding is over, at the beginning of fruit ripening, caterpillars move down the plant and gnaw into the stem for pupation. More than ten pupae may be found within a single stem. Adults appear after 3–4 weeks in July and some time later they seek for overwintering shelters (Wêgorek 1966). This is a common pest, especially in the regions of caraway intensive cultivation. Its mass appearance could be avoided by an adequate crop rotation with a limited participation of caraway. However, protection against this pest is mainly based on chemical control. The main spraying should be performed in spring of the second growing season, upon monitoring conducted over the initial period of adults' appearance. During their mass flight the treatment should be repeated at least 3–4 days prior to flowering. If the pest is abundant during flowering, its chemical control can be carried out only in the evenings for bee prevention. At initial infestation by young larvae, caraway can be treated with biological formulations (e.g. Dipel 3.2 WP, Bactospeine PM 6000, Bacilan). Chemical control against this species, as against other pests and diseases, should be conducted upon the actual recommendations established in particular countries.

Another serious pest of caraway crop is *Aceria carvi* Nal., invisible by the naked eye. Adult mites overwinter within caraway leaf rosettes and in March they start feeding. In April the females deposit eggs on the surface of leaf blades or at their base. The nymphs which hatch from eggs, after two weeks transform into adults. Several generations may develop during the growing season. In September the mites move towards overwintering shelters. Caraway is attacked by this pest in both years of vegetation. Heavily infested rosette leaves turn curly and malformed, then even wilt. In the second year the mites feed both on leaves and flowers. Infested foliage is covered with light spots, whereas the flowers become malformed, changing colour to pink, lilac or greenish. Their stamens and pistils are shapeless, resembling the petals. The mites develop within receptacle, thus the flowers degenerate. In consequence, the fruits may not set, but if so, they are minute and malformed. In the regions where this mite is abundant chemical control should be carried out. In the first growing season a double spray (in summer and early autumn) is required, whereas next year at least one treatment should be performed at the beginning of vegetation (Königsmann 1958, Plescher 1989, Pruszyński 1995).

Caraway crops are often infested with aphids. Leaves, shoots and umbels are usually populated by two: willow-carrot aphid (*Cavariella aegopodii* Scop.) and hawthorn-carrot aphid (*Dysaphis crategi* Kalt.). Also, root aphids, attacking most umbelliferous plants, can feed on caraway underground organs.

Chemical control should be conducted when the first aphid colonies appear on the crop. In the case of heavy infestation, the treatment should be repeatedly performed roughly at 10-day intervals. Some of insecticides recommended against *Depresaria nervosa* caterpillars can also effectively suppress aphid populations (Ester *et al.* 1993, Pruszyński 1995).

Likewise other umbelliferous plants, caraway can be attacked by capsids, especially *Lygus campestris* L. and *Lygus kalmi* L. Damage is caused by both larvae and adults. Heavily infested plants turn yellow, then brown and finally dry up. Apart of direst harmless, capsids are known as carriers of various diseases. Chemical treatments against these pests are similar to those controlling the aphids.

Caraway crops can be also considerably affected by polyphagous pests such as grubs (*Melolonthidae*) and wireworm (*Elateridae*). If abundant, these pests should be eliminated with insecticides mixed with the soil in the year preceding a setting up of a plantation. Occasionally, caraway crop can be also invaded by rodents, particularly the vole (*Microtus arvalis* Pall.), field mouse (*Apodemus agrarius* Pall.) and wood mouse

(*Apodemus sylvaticus* L.). All these pests gnaw the young plants and devour the roots during caraway winter dormancy. Rodents are difficult to be controlled, sometimes they are treated with exhaust fumes.

Caraway plants can be infected by numerous diseases (Ondrej 1983), however only a few of them are of a major importance. Elimination of seed-born diseases, especially those causing seedling damping-off, seems very essential. A quite efficient preventive measure is fungicidal seed-dressing just before sowing. During the vegetation season, caraway is often infected by downy mildew (*Plasmopara nivea* Schort.) which if intensive, can cause yellowing and wilting of the plants. Therefore, this pathogen should be treated with fungicides. Caraway stems can be attacked by basal root (*Phoma anethi* Sats.) and grey mould (*Botrytis cinerea* Pers.).

So called gangrene of inflorescence, caused by bacteria such as *Erwinia*, *Pseudomonas* and *Xanthomonas*, often appears while umbels are being formed and during flowering. In the Netherlands the most serious diseases of caraway are considered anthracnose (*Mycocentrospora acerina*) and white mould (*Sclerotinia sclerotiorum*). The first pathogen can be considerably limited by particular practices (e.g. lowering of N-fertilization and sowing rate), chemical treatments, and by biological control (Evenhuis and Verdam 1995, Verdam *et al.* 1993).

REFERENCES

Aflatuni, A., Palevitch, D. and Putievsky, E. (1993) The effect of manure composted with drum composter on aromatic plants. *Acta Horticulturae*, **344**, 63–68.

Bernáth, J. (1993) Vadon termö és termesztett gyógynövények, *Mczögazda Kiadó*, Budapest. pp. 179–183.

Bocheñska, I. and Kozlowski, J. (1969) Badania wahañ zawartosci olejku, wilgotnosci oraz zdolnosci kielkowania owoców kminku zwyczajnego (*Carum carvi* L.) w czasie dojrzewania i przechowywania. *Herba Polonica*, **15**, 251–253.

Boshart, K. (1942) Über Anbau und Düngung aromatischer Pflanzen. *Heil- u. Gewürzpflanzen*, **21**, 73–91.

Buszczak, T. (1962) Studia nad kminkiem zwyczajnym (*Carum carvi* L.). Cz.II Wplyw wilgotnosci gleby na rozwój i strukturê plonu. *Pamiêtnik Pulawski*, **7**, 137–156.

Chládek, M. (1969) Modifikace mikrofenologické metody pri kontrole vzrostného vrcholu (*Carum carvi* L.). *Genetika a Slechteni*, **3**, 185–192.

Chládek, M. (1974) Pozadawek jarovizace u rostlin *Carum carvi* L. *Genetika a Slechteni*, **10**, 155–162.

Chotin, A.A. (1959) O stadijnom razvitii nikotorych efiromasliènych rastìnij. *Agrobiologia*, **116**, 231–235.

Chotin, A.A. and Szulgina, G. (1963) *Efiromaslicnyje kultury*, ISLZiP, Moskva.

Czabajski, T., Golcz, L. and Jaruszewski, W. (1960) Wplyw nawozów mineralnych na plon i zawartosc olejku u kminku zwyczajnego (*Carum carvi* L.). *Biuletyn Instytutu Roslin Leczniczych*, **6**, 89–95.

Ester, A., Vreeke, S., Floot, H.W.G. and Wander, J.G.N. (1993) Control of caraway root aphid in caraway. *Mededelingen van de Faculteit Landbouwwetenschappen*, **58**, 653–659.

Evenhuis, A. and Verdam, B. (1995) Possibilities to control anthracnose of caraway in order to increase yield stability. Results from 1990–1994. *Verslag Proefstation voor de Akkerbouw en de Groenteteelt in de Vollegrond*, **189**, 166.

Hecht, H., Mohr, T. and Lembrecht, S. (1992) Harvesting medicinal grains by combine. *Landtechnik*, **47**, 494–496.

Heeger, E.F. (1956) Kümmel (*Carum carvi* L.). *Handbuch des Arznei -u. Gewürzpflanzenbaues*. Berlin, 328–338.

Jankulov, J.K. (1960) O faze razwitija pri kotoroj kmin prochodit stadiju jarovizacji. *Doklady Bolgarskoj Akademii Nauk.* **13**.

Jaruzelski, M. (1974) Kminek zwyczajny (*Carum carvi L.)* – znaczenie i ukierunkowanie upraw wspólrzędnych. *Wiadomosci Zielarskie*, **3**, 1–2.

Kordana, S., Lesniewska, S. and Golcz, L. (1983) Potrzeby pokarmowe kminku zwyczajnego (*Carum carvi L.*) *Herba Polonica*, **29**, 27–38.

Kordana, S. and Zalęcki, R. (1993) Dolistne dokarmianie kminku zwyczajnego (*Carum carvi L.*). *Herba Polonica*, **39**, 197–203.

Königsmann, E. (1958) Untersuchungen an der Kümmelgallmilbe (*Aceria carvi Nal.*). Wiss. Z. Karl–Marx Univ. Leipzig, *Mat. Nat. Reihe*, **7**, 329–349.

Kramer, W. (1955) Ein Beitrag zum Kümmelanbau. *Die Pharmazie*, **10**, 550–554.

Kubis, A. (1973) Uprawa kminku wspólrzędnie z lnem. *Zielarski Biuletyn Informacyjny*, **11**, 6–7.

Mihalea, A. (1965) Influenta ingrasaminterol chimice asupra chimionulini (*Carum carvi L.*). Cultivat in sudestul transilvaniei. *Probleme agricole*, **1**, 25–27.

Müller, H.R., Pank, F. and Plescher, A. (1989) Anbauverfahren Kümmel (*Carum carvi L.*). 1. Mitteilung, *Drogen Report*, **3**, 77–86.

Müller, H.R. (1990) Anbauverfahren Kümmel (*Carum carvi L.*). 2. Mitteilung, *Drogen Report*, **4**, 35–45.

Nordestgaard, A. (1986) Growing caraway (*Carum carvi L.*) for seed. Sowing and nitrogen rates. *Tidsskrift for Planteavl*, **1**, 37–44.

Nowiński, M. (1959) *Rosliny lecznicze flory polskiej*. PWN, Poznań.

Novák, V. (1973) Essence of conditional perennity of caraway and possibilities of use the respective investigations for widening its production. *Roczniki Nauk Rolniczych*, **99**, 11–24.

Ondrej, M. (1983) The occurrence of fungi on caraway (*Carum carvi L.*) in Czechoslovakia. *Sbornik-UVTIZ, Ochrana Rostlin*, **19**, 235–237.

Plescher, A. (1989) Bekämpfung tierischer Schaderreger im Kümmelanbau. *Drogen Report*, **3**, 29–38.

Pruszyński, S. (1995) *Zalecenia ochrony roslin na lata 1995/96*. IOR, Poznań.

Rumińska, A. and Kaczor, Z. (1963) Wplyw wapnowania na wzrost i plon kminku zwyczajnego (*Carum carvi L.*). *Zeszyty Naukowe SGGW, Rolnictwo*, **7**, 99–113.

Rumińska, A. (1981) *Rosliny lecznicze*. PWN, Warszawa.

Rumińska, A. (1990) *Poradnik plantatora ziól*. PWRiL, Poznań.

Scharrer, K. and Lisner, H. (1965) *Handbuch der Pflanzenernährung und Düngung*. Springer-Verlag, Wien–New York.

Schröder, H. (1964) *Arznei- und Gewürzpflanzen*. Bernburg/Saale.

Węglarz, Z. (1982) Wplyw czynników agrotechnicznych na przechodzenie kminku zwyczajnego (*Carum carvi L.*) z fazy wegetatywnej w generatywna. Cz.1. Wplyw wielkosci wysadków na wartosc nasienników kminku zwyczajnego. *Herba Polonica*, **28**, 171–177.

Węglarz, Z. (1983/a) Wplyw czynników agrotechnicznych na przechodzenie kminku zwyczajnego (*Carum carvi L.*) z fazy wegetatywnej w generatywna. II. Wplyw nawozenia i wilgotnosci gleby na rozwój i plonowanie kminku zwyczajnego. *Herba Polonica*, **29**, 21–26.

Węglarz, Z. (1983/b) Wplyw czynników agrotechnicznych na przechodzenie kminku zwyczajnego zwyczajnego (*Carum carvi L.*) z fazy wegetatywnej w generatywna. III. Wplyw terminu siewu, ilosci wysiewu nasion i poziomu nawozenia na rozwój i plonowanie kminku zwyczajnego. *Herba Polonica*, **29**, 103–111.

Węgorek, W. (1966) *Nauka o szkodnikach roslin*. PWRiL, Warszawa, 265–266.

Verdam, B., Gerlagh, M. and Van-de-Geijn, H.M. (1993) Biological control of *Sclerotinia sclerotiorum* in caraway (*Carum carvi L.*). *Mededelingen van de Faculteit Landbouwwetenschappen*, **58**, 1343–1347.

Vrzalowa, E. (1955) Vliv predsetove upravy osiva kminu na jeho rychlest kliceni. *Sbornik Vysoke Skoly Zemledelske a Lesnicke Fakulty*. Brno **3**, 244–252

Ziemlinskij, E. (1954) *Lekarstwiennyje rastienija SSSR*, Moskva.

9. AGROTECHNOLOGY OF ANNUAL CARAWAY PRODUCTION

ELI PUTIEVSKY

Division of Aromatic Plants, Agricultural Research Organization,
Newe Ya'ar Research Center, P.O. Box 1021, Ramat Yishay 30095, Israel

9.1. INTRODUCTION

The annual form of caraway (*Carum carvi* L.) is grown commercially in temperate zones (Mediterranean, Africa, S. America and Asia), while the biennial form is grown mainly in the northern parts of the world (Europe and America). As the seed is the only part of the plant used, the quality of the biennial form is considered to be superior to the annual one, according to size, aroma, colour etc. (Bouwmeester 1991). The geographic distribution and quality differences encourage in the last years the development of an annual form suitable to grow in the northern part of the world (Bouwmeester and Kuijpers 1993, Kallio *et al.* 1994), as well as the selection of an annual form out of the biennial types suitable to grow in the temperate zones (Nordestgaard 1986). As caraway has become very important, not only for traditional purposes, but as a source for carvone, used commercially as a sprouting inhibitor for potatoes, many researches are done in order to increase the level of carvone (Bouwmeester *et al.* 1995a, Kallio *et al.* 1994, Mheen 1994, Toxopeus and Lubberts 1994, Toxopeus *et al.* 1995), including enantiomers of carvone appearing in different varieties (Bouwmeester *et al.* 1995a) and practical agrotechnological methods, as well as harvest and postharvest methods (Anonymous 1988, Hecht *et al.* 1992, Singh *et al.* 1992, Wander and Zwanepol 1994).

This chapter will review the information available in the literature about different agrotechnology elements affecting the production of annual caraway.

9.2. VARIETIES AND SELECTION

In the Mediterranean region varieties that seem to originate from local wild populations, or their origin is unknown, are called in Arabic "Balady". These 'types' became the source of variation for selection of new types (=varieties?). In Egypt, for example, they select an annual caraway variety for essential oil production (called No. 1-4-36) from the Balady (El-Ballal 1979), and continue the selection of progenies by breeding and isolation (El-Ballal 1978, Munshi 1979). The main breeding, selection and introduction had been made in eastern European countries (mainly Hungary, Poland, East Germany, Czechoslovakia, etc.) (Prochazka and Urbanova 1972, Pushmann *et al.* 1992), which sometimes had genetic projects with other countries like Egypt (Chladek *et al.* 1974, Köck 1987). During the last years researchers in the Netherlands started a selection and breeding program with annual caraway (Hälvä *et al.* 1986, Kordana *et al.* 1983), including non-shattering varieties (Anonymous 1988 and 1992), while in other countries, like Egypt and Israel, an effort had been made to grow the biennial types (Chladek *et al.* 1974, Putievsky 1993, Putievsky *et al.* 1994).

From a genetic point, only few works have been done on DNA count (Das 1991) or chromosome numbers and chiasma distribution of annual caraway (Sheidai *et al.* 1996). Doubling the chromosome numbers (autotetraploid) decrease the seed production but increase the essential oil content and the carvone (Dijkstra and Speckmann 1980, Zderkiewicz 1971).

In order to get the highest seed, essential oil and carvone yield, the quickest way is to examine at the growing sites as many varieties and cultivars as possible and to use breeding programs for improvement of yield (Dusek 1992), but by crossing a biennial variety (Bleija) and an annual variety (Karzo), researchers from the Netherlands produced an annual population with yielded about 20% higher carvone. To preserve this population they used an *in vitro* propagation method (Toxopeus *et al.* 1995).

9.3. CULTIVATION

9.3.1. Germination Characteristics

The phenomenon of seed dormancy is well known in species belonging to the Umbelliferae family, especially in caraway (Putievsky 1977, 1979, 1980). The dormancy can be caused by all or by some of three factors – action of endogenous aromatic substances, deficiency of endogenous growth promoter/s or by endogenous inhibitors (Hradilik and Cisarova 1975).

Running water during a few days found to increase germination rate and level, may be by removing the endogenous inhibitors (Putievsky 1977, 1980), but cold treatment before sowing (Hradilik and Cisarova 1975) or at germination stage (Putievsky 1980) also increased germination. Dormancy caused by deficiency of endogenous growth promoters can be treated by adding artificial gibberellins like GA, ABA or cytokinin, but cold stratification reduces up to zero the need for growth promoters (Hradilik and Cisarova 1975, Hradilik and Fiserova 1980).

The optimal time of sowing annual caraway is during the autumn (October–November) when the temperature is moderate in the daytime (15–22°C) and quite cold during the night (7–13°C). In control germination experiments, the optimal temperature has been found to be 15°C/10°C (day/night) (Putievsky 1983), (Figure 1), such conditions exist in the fields in the autumn. Therefore, at Mediterranean regions where annual caraway is growing, it should not be sown in mid winter, when temperatures are extremely low, unfavourable for germination as well as for rapid seedling development. The fact that germination under field conditions takes a relatively long time can be explained by the amount of inhibitors existing in the seeds, which have to be washed out by artificial irrigation or by rain. Cold nights at the germination stage can solve the deficiency of growth promoters. In modern mechanized agriculture it is usually essential to achieve rapid uniform germination and to minimize the time between sowing and

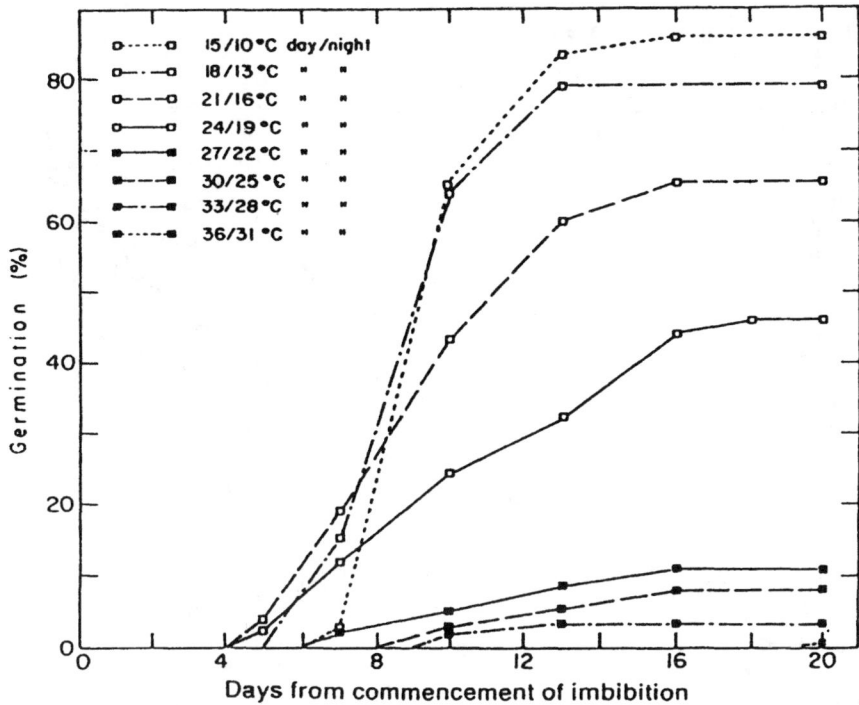

Figure 1 Effect of temperature on germination of caraway seed

the emergence of the seedling. Such considerations are particularly important in arid regions where irrigation is costly or even impossible. Therefore it is of great importance to use only normal seeds with embryos (Warakomska 1989) and to find chemicals that can increase the germination ratio (Benjamini 1986).

Leaching as a pre-treatment in the laboratory (Lott *et al.* 1991), rather than in the field, minimized the interval between planting and seedling emergence (by more than 50%) and increased the level of germination (Figure 2), (Putievsky 1983). Such treatments were most effective where there was a saving of six days and germination was increased by 13% relative to the controls.

9.3.2. Sowing and Spacing

Sowing date, and depth, are the most important conditions for field establishment. Considering the seed size of caraway (1–3 g/1000 seeds), the sowing should not be deeper than 1.5–2.0 cm, while the optimal depth is around 1.0 cm (Hornok 1986, Putievsky 1977). The main problem is that during the germination period the upper soil (1–2 cm) should be continuously wet, otherwise the seeds will lose their ability to germinate. Sowing deeper than 2 cm can solve the dependence on rain intervals, but that can be done only with varieties that have big seeds (above 3 g/1000 seeds). Biennial caraway is usually sown during the raining season (spring), therefore the sowing depth is

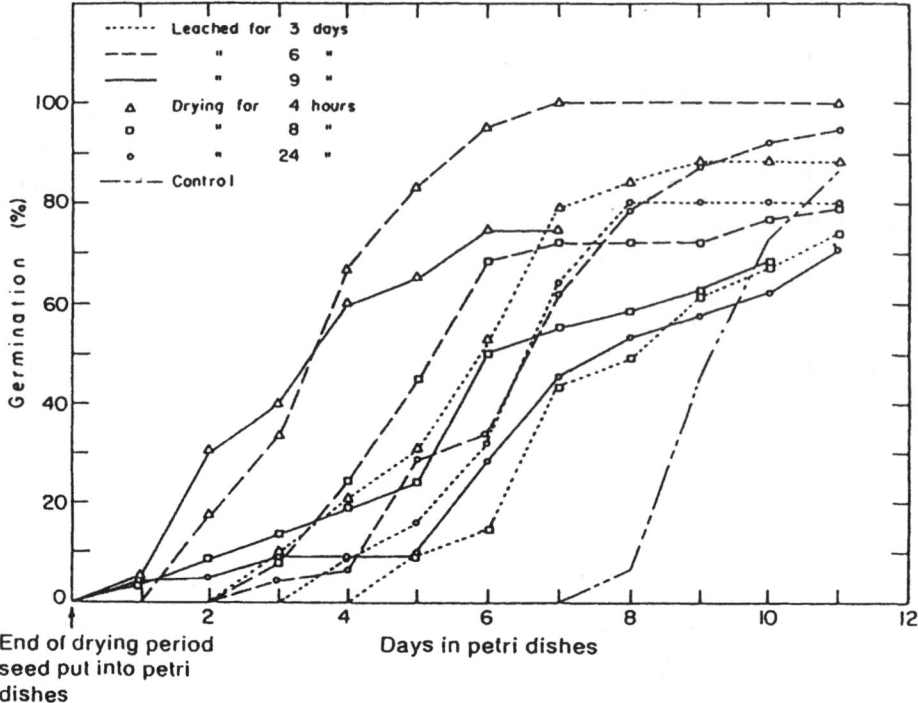

Figure 2   Effect of leaching and drying on subsequent germination of caraway seed

less important, but in the Mediterranean, where annual caraway is sown in autumn, it is dangerous to rely only on the rain.

It was found that for maximum seed production per plant, the living space of the plant should enable it to express its maximum potential, i.e. that plant has maximum light, water, fertilization, etc. (Hornok 1986). The experiment, that was done under Mediterranean conditions, supports this statement (Figure 3) (Putievsky 1976). As can be seen, increasing plant density decreases the seed production per plant, while under field conditions, the maximum seed yield is obtained with 60 plants/m^2 (Figure 3). To obtain such plant density, it was found that 6 kg/ha of seeds should be sown (Nordestgaard 1986, Putievsky 1976, Putievsky and Kuris 1977). In other countries (e.g. climate and soil condition) and with other varieties, plant density was found to be different (Hornok and Csáki 1982, Munshi et al. 1990).

9.3.3. Growth Regulators

Most of the work on growth regulators, examining the effects on yield components including essential oil, were done by Egyptian researchers. It was found that CCC (chlormequat) (at 250–1000 ppm) applied to young plants even decreased the plant height compared to GA3 or IAA, but increased the oil yield (Sarhan and El-Sayed 1983). The same effect on plant height (decrease) was found with daminozide

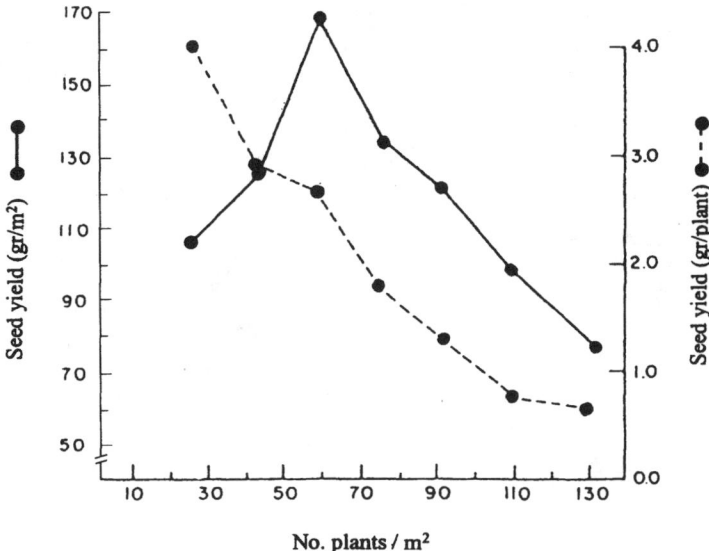

Figure 3   The effect of plant number on caraway yield

(at 500–4000 ppm), but it increased umbel number, seed weight, essential oil and fat in the seeds (Abou Zied 1974). In other experiments made with GA3 (at 50–200 ppm) it reduced seed and essential oil production but the plant height increased (Ahmed and Eid 1975), while with CCC they found the same results as before (Sarhan and El-Sayed 1983). In order to use those results in a commercial way, it is necessary to examine them in open and commercial fields.

9.3.4. Fertilization

Most of the researches on fertilization test only different levels of different fertilizers and their effect on yield components. No data on soil elements content at the start or at the end of the experiments or the relationship between results and type of soil are presented. Eventhough, there is some evidence that fertilization is very important for plant growth as well as for seed production at different climates (Dachler 1990, El-Gamal *et al.* 1983, Kordana *et al.* 1983, Lihan and Jezikowa 1991, Munshi *et al.* 1990, Nordestgaard 1986, Putievsky 1976, Rieder and Reiner 1972, Weglarz 1983).

In Europe it was found that N is needed mainly during leaf development and K during flowerstalk growth while the P and Ca uptake (as found in the plant parts) was high during seed ripening (Kordana *et al.* 1983, Lihan and Jezikowa 1991, Nordestgaard 1986, Rieder and Reiner 1972, Weglarz 1983). Only in one report (Dachler 1990) nitrogen didn't have any influence on caraway growing in Europe.

Annual caraway responses very positively to N, P, i.e. plant height, number of branches, seed weight and seed yield (El-Gamal *et al.* 1983, Munshi *et al.* 1990, Putievsky 1976, Putievsky and Sanderovich 1985). No effect has been found on essential oil content and composition.

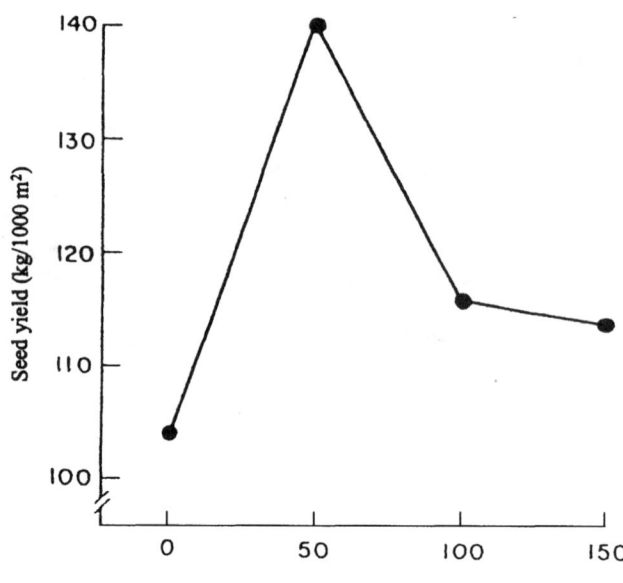

Figure 4   The effect of nitrogen fertilization on caraway seed yield

One of the most important problems of fertilization is the time of application: should it be applied only before sowing or also during the growing period? The highest seed production has been obtained when high level of N is applied before sowing or 50% before sowing and 50% of the total amount at mid winter (Putievsky 1976, Putievsky and Kuris 1977, Putievsky and Sanderovich 1985). Different N sources, like calcium nitrate, ammonium sulphate or urea, don't have any effect on yield components (El-Gamal *et al.* 1983).

In Israel we found that the maximum seed yield has been obtained at 50 kg of $N/1000\,m^2$ supplied as ammonium sulphate (Figure 4) (Putievsky and Kuris 1977, Putievsky and Sanderovich 1985).

9.3.5. Irrigation

In semiarid regions where the annual caraway grows, there are two critical stages during its growth, where irrigation is necessary: from germination to establishment and at seed formation (Hornok 1986, Penka 1978, Weglarz 1983). The first stage appears at autumn and the other at spring/early summer (Putievsky 1976). At those periods the amount and date of rainfall change from year to year, and in most of the places, where the annual caraway grows, it is impossible to irrigate artificially. Furthermore, as this plant is considered a winter crop, grown commercially on a large scale, the farmer cannot search for fields close to a source of water that can be used for artificial irrigation (Putievsky *et al.* 1984, Putievsky and Kuris 1977, Putievsky and Sanderovich 1985).

In Egypt, when rainfall is not sufficient, the farmer makes use of the flooding system to irrigate his crop, while in Israel in commercial fields most of the farmers have the possibility, in extreme conditions, to use sprinkler irrigation.

9.3.6. Weed and Disease Control

Weed control is a very important factor mainly during the early developing stages before plants cover the field, at early spring (March). During this period the ground is bare and winter weeds can grow very easily. From early spring the caraway plants cover the ground and normally (when there are enough plants per unit area) no addition of annual weeds can be observed. The growing season of biennial caraway is much longer, therefore a wide range of weed control is needed for a longer time. Most of the weed controls found suitable for the establishment phase of biennial caraway (Hajdú and Földesi 1972, Mahadeva and Reesor 1973, Mitchell and Abernethy 1993, Pank *et al.* 1984, Pinzaru 1983), are also suitable for the annual caraway (Putievsky 1976). The two most common chemicals used for weed control, Afalon (linuron) and prometryne, are mainly used after sowing before emergence (Mahadeva and Reesor 1973, Mitchell and Abernethy 1993, Pinzaru 1983, Putievsky 1976), but can be used also before sowing (Hajdú and Földesi 1972, Pank *et al.* 1984). Those two herbicides can be used also as post-emergence, beside chlorbromuron, nitrofen or aresin (monolinuron). No direct effect was found of the weed control on the yield components, beside the fact that when there are no weeds there is no competition and the cultivated caraway plants can grow without any interference. In Israel, just before the soil preparation 2.0 kg/ha of trifluralin is sprayed and mixed with the soil, in order to prevent germination of perennial weeds, like purple nutsedge (*Cyperus rotundus*), Johnsongrass (*Sorghum halepense*) and Field bindweed (*Convolvulus arvensis*). We have found that the use of most of the weed controls against annual cereals (sethoxyd, fluazifop-buthyl or dethodin) didn't do any damage to caraway and can be used as post-emergence (Pank *et al.* 1984).

In biennial caraway parasitic fungi were found in all plant parts (Ondrej 1983) as well as different pests and diseases (El-Sayed *et al.* 1990, Evenhuis and Verdam 1995, Hluchy 1985, Plescher and Herold 1983) but no information on such problems has been reported in literature on annual caraway. On the other hand, some fungi (*Aspergillus flavus*, *A. niger* or *Fusarium moniliforme*) had been found in annual caraway seeds, with negative effect on its quality (Regima and Tulasi 1992). If harvest and post-harvest processes are done correctly, there is no reason to believe that the seeds should be damaged.

9.3.7. Physiology and Yield

Seed yields of caraway vary considerably between years at the same site and in the same year at different sites with the same variety (Ben Yehuda and Putievsky 1985). It means, that environmental conditions affected the yield or determining processes of seed production (Bouwmeester and Smid 1995, Prochazka and Urbanova 1972, Toxopeus and Lubberts 1994).

In order to understand the life cycle of annual caraway under Israeli conditions, the ratio between different plant parts (Figure 5) and at different stages (in connection with

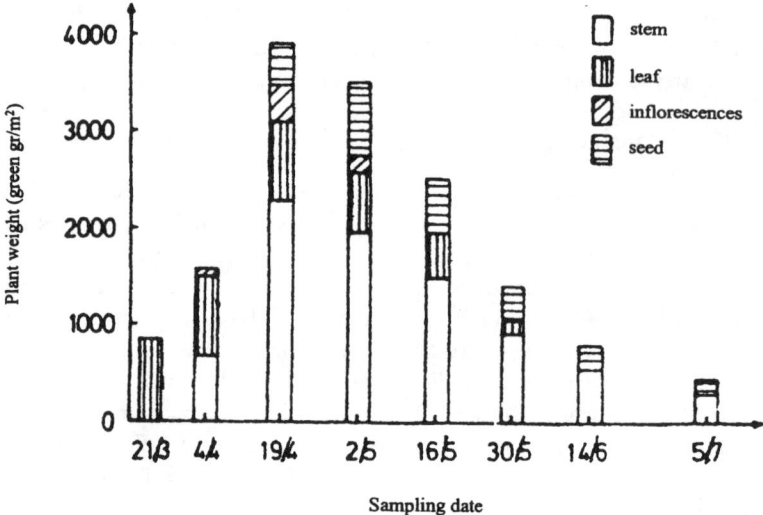

Figure 5   The weight of caraway plant parts at different dates

Table 1   The effect of phenological stage and storage on essential
oil yield and components of annual caraway

Sampling date	Phenological stage	Essential oil yield (cc/m^2)	Main constituents in the essential oil (%)	
			Limonene	Carvone
4.4	Flowering initiation	0.9	—	—
2.5	Flowering	8.8	—	—
16.5	"Milk" stage	12.1	47	48
30.5	Wax stage	5.6	32	58
14.6	Dry plant	4.2	31	63
5.7	Seed collection	2.7	19	78
11.11	Seed storage	2.5	38	52

the dates) (Figure 5), the essential oil yield and composition were examined (Table 1) (Putievsky *et al.* 1984). While the maximum essential oil yield was obtained during "milk" stage ($12\,cc/m^2$) the best composition (high carvone content) appeared at harvest time (78%) (Fleischer *et al.* 1988, Witchmann and Stahl-Biskup 1987, Zderkiewich 1971). Storage of the seed for a few months reduced the carvone content (Fleischer *et al.* 1988, Putievsky *et al.* 1984).

The question, if to use the plants for seed or for essential oil production still depends on the prices of each on the international market (Basket and Putievsky 1976).

Light intensity, day length and light accumulation seem to be important factors, mainly during flowering stage and early stage of seed development, affecting quantity and quality of seed and essential oil production. These factors affect the photosynthetic activity and therefore the assimilation availability (Bouwmeester *et al.* 1995b,c, Ceska *et al.* 1987, Putievsky 1983, Smid and Bouwmeester 1993). It seems that in climates, where the annual caraway grows, light is not a limiting factor.

Pollination also affects seed size and yield. Though some of the pollination is done by the wind (Bouwmeester and Smid 1995, Smid and Bouwmeeseter 1993, Toxopeus and Bouwmeester 1992) the most important and effective pollinators are bees, but also some other insects (El Berry *et al.* 1974, Hussein *et al.* 1991, Omar *et al.* 1991, Ricciardelli d'Arbore 1983). The time of insect activity of some species is also investigated and varies between 11 a.m. and 4 p.m. (47).

Undoubtedly, weather conditions, i.e. rain, humidity and wind also affect pollination efficiency (Omar *et al.* 1991, Bouwmeester and Smid 1995).

REFERENCES

Abou Zied, E.N. (1974) Increase in volatile oil and chemical composition in the seeds of caraway and fennel plants induced by succinic acid-2,2-dimethylhydrazide. *Biol. Plant.*, **16**, 123–126.

Ahmed, S.S. and Eid, M.N.A. (1975) Effect of gibberellic acid and Cycocel on yield of seeds and essential oil of some umbelliferous plants. *Egypt. J. Hortic.*, 2(2), 227–232.

Anonymous (1988) *Stichting voor Plantenveredeling-SVP. Caraway.* Jaarverslag 1987, **68**, 120 (in Dutch).

Anonymous (1992) *The 68th list of approved crop varieties, 1993.* Commissie voor de Samenstelling van de Rassenlijst voor Landbouwgewassen. Centrum voor Plantenveredelings- en Reproduktieonderzoek, Wageningen, The Netherlands (in Dutch).

Avtar Singh, Mahey, R.K. and Singh, A. (1992) Production technology of caraway [*Carum carvi*], cumin [*Cuminum cyminum*] and ajowan [*Trachyspermum ammi*] – a review. *Crop Res. Hisar*, **5**, Suppl., 1–10.

Basker, D. and Putievsky, E. (1976) Essential oils of *Carum carvi. Hassadeh*, **57**, 209, 211–212 (in Hebrew).

Ben Yehuda, R. and Putievsky, E. (1985) Cultivation of herbs and spices in the Golan Heights. *Hassadeh*, **65**, 1358–1359 (in Hebrew).

Benjamini, L. (1986) Effect of carbofuran on seed germination and initial development of seven crops. *Phytoparasitica*, **14**, 219–230.

Bouwmeester, H.J. (1991) *Production of caraway essential oil. A study of the literature.* Verslag Centrum voor Agrobiologisch Onderzoek, 1991, No. **150** (in Dutch).

Bouwmeester, H.J. and Kuijpers, A.M. (1993) Relationship between assimilate supply and essential oil accumulation in annual and biennial caraway (*Carum carvi* L.). *J. Ess. Oil Res.*, **5**, 143–152.

Bouwmeester, H.J. and Smid, H.G. (1995). Seed yield in caraway (*Carum carvi*). 1. Role of pollination. *J. Agric. Sci.*, **124**, 235–244.

Bouwmeester, H.J., Davies, J.A.R. and Toxopeus, H. (1995a) Enantiomeric composition of carvone, limonene, carveols in seeds of dill and annual and biennial caraway varieties. *J. Agric. Food Chem.*, **43**, 3057–3064.

Bouwmeester, H.J., Smid, H.G. and Loman, E. (1995b) Seed yield in caraway (*Carum carvi*). 2. Role of assimilate availability. *J. Agric. Sci.*, **124**, 245–251.

Bouwmeester, H.J., Davies, J.A.R., Smid, H.G. and Welten, R.S.A. (1995c) Physiological limitations to carvone yield in caraway (*Carum carvi* L.). *Industrial Crops and Products*, **4**, 39–51.

Ceska, O., Chaudhary, S.K., Warrington, P.J. and Ashwood-Smith, M.J. (1987) Photoactive furo-coumarins in fruits of some umbellifers. *Phytochemistry*, **26**, 165–169.

Chladek, M., Machovicova, F., Tyllova, M. and Mesarosova, L. (1974) Investigations on the essential oil quality of annual caraway (*Carum carvi* f: *annuum*) by thin layer chromatography. *Bull. Vyzkumny Ustav Zelinarsky Olomouc*, No. **18**, 73–81 (in Czech).

Dachler, M. (1990) Varieties and nitrogen requirements of some medicinal and spice plants grown for seeds (*Papaver somniferum* L., *Linum usitatissimum* L., *Carum carvi* L. and *Sinapis alba* L.). Int. Symp. on Medicinal and Aromatic Plants, Budapest, Hungary, 1990. *Herba Hung.*, **29**, 41–50.

Das, A.B. (1991) Genome analysis and variation of 4C DNA content in the subtribe Carinae. *Cytologia*, **56**, 627–632.

Dijkstra, H. and Speckmann, G.J. (1980) Autotetraploidy in caraway (*Carum carvi* L.) for the increase of the aetheric oil content of the seed. *Euphytica*, **29**, 89–96.

Dusek, K. (1992) Comparison of various caraway genotypes in Czechoslovakia. *Zahradnictvi*, **19**, 151–160 (in Czech).

El-Ballal, A.S.I. (1978) Stability of selection response for high essential oil yield in local caraway (*Carum carvi* L.). *Acta Horticulturae*, No. **73**, 59–64.

El-Ballal, A.S.I. (1979) Genotype-environmental interaction in essential oil yield in the selected caraway type 1.4.36. *Herba Hung.*, **18**, 155–166.

El-Berry, A.R., Gawaad, A.A.A., Moustrafa, M.A.F. and El-Gayar, F.H. (1974) Pollinators other than honey bees visiting certain medicinal plants in Egypt. *Z. Angew. Entomol.*, **76**, 113–119.

El-Gamal, E.A., Mahmoud, M.M. and Omar, F.A. (1983) Effect of some fertilizer treatments on yield and volatile oil content of *Carum carvi* L. *Ann. Agric. Sci.* (Cairo), **28**, 1605–1614.

El-Sayed, A.M.K., Abdel-Galil, F.A., Darwish, Y.A., El-Hagag, G.H.A. and Abou-El-Hagag, G.H. (1990) Incidence and dominance of arthropods associated with roselle, caraway and coriander plants in Upper Egypt. *Assiut J. Agric. Sci.*, **21**, 153–165.

Evenhuis, A. and Verdam, B. (1995) *Possibilities to control anthracnose of caraway in order to increase yield stability. Results from 1990–1994.* Verslag Proefstation voor de Akkerbouw en de Groenteteelt in de Vollegrond, 1995, No. **189** (in Dutch).

Fleisher, A., Fleisher, Z., Lawrence, B.M. (ed.), Mookherjee, B.D. (ed.) and Willis, B.J. (ed.) (1988) The essential oil of annual *Carum carvi* L. grown in Israel. In: Flavors and Fragrances: a world perspective, Proc. 10th Int. Congress of Essential Oils, Fragrances and Flavors, Washington, DC, USA, 1986. *Dev. Food Sci.*, **18**, 33–40.

Hajdu, K. and Földesi, D. (1972) Chemical weed control in caraway (*Carum carvi* L.). *Herba Hung.*, **11**, 45–51 (in Hungarian).

Hälvä, S., Hirvi, T., Makinen, S. and Honkanen, E. (1986). Yield and glucosinulate of mustard seeds and volatile oils of caraway seeds and coriander fruit. II. Yield and volatile oils of caraway seeds (*Carum carvi* L.). *J. Agric. Sci.* (Finland), **58**, 163–167.

Hecht, H., Mohr, T. and Lembrecht, S. (1992) Harvesting medicinal grains by combine. *Landtechnik*, **47**, 494–496, 504 (in German).

Hluchy, M. (1985) A mass outbreak of the tortricid Cnephasia virgaureana on caraway. Sb. UVTIZ (Ustav Vedeckotech. Inf. Zemed.) *Ochr. Rostl.*, **21**, 261–266 (in Czech).

Hornok, L. (1986) Effect of environmental factors on growth, yield and on the active principles of some spice plants. *Acta Horticulturae*, No. **188**, 169–176.

Hornok, L. and Csáki, G. (1982) Effect of stand density on caraway (*Carum carvi* L.). *Herba Hung.*, **21**, 59–65 (in Hungarian).

Hradilik, J. and Cisarova, H. (1975) Studies on the dormancy of caraway (*Carum carvi*) achenes. *Rostlinna Vyroba*, **21**, 351–364 (in Czech).

Hradilik, J. and Fiserova, H. (1980) The role of abscisic acid in caraway (*Carum carvi*) seed dormancy. *Acta Univ. Agric. Brno A Fac. Agron.*, **28**, 39–64 (in Czech).

Hussein, M.H., Omar, M.O.M., Darwish, Y.A. and Abdalla, M.A. (1991) Effect of insect pollination and quality of cumin, caraway and anise seeds in Assiut region. *Assiut J. Agric. Sci.*, **22**, 69–79.

Kallio, H., Kerrola, K. and Alhonmaki, P. (1994) Carvone and limonene in caraway fruits (*Carum carvi* L.) analyzed by supercritical carbon dioxide extraction-gas chromatography. *J. Agric. Food Chem.*, **42**, 2478–2485.

Köck, O. (1987). SzK-1 caraway cultivar. *Kertészet és Szölészet*, **36**, 8 (in Hungarian).

Kordana, S., Lesniewska, S. and Golcz, L. (1983) Nutritional requirements of caraway (*Carum carvi* L.). *Herba Pol.*, **29**, 27–38 (in Polish).

Lihan, E. and Jezikova, O. (1991) Long-term effect of the nutrition of grassland coenoses. *Dlhodoby Ucinok Vyzivy Pratocenoz.* Vedecke-Prace-Vyskumneho-Ustavu-Luk-a-Pasienkov-v-Banskej-Bystrici., **21**, 63–73.

Lott, J.N., Cavdek, V. and Carson, J. (1991) Leakage of K, Mg, Cl, Ca and Mn from imbibing seeds, grains and isolated seed parts. *Seed Sci. Res.*, **1**, 229–233.

Mahadeva, S. and Reesor, R.A. (1973) Herbicidal trials on caraway. *Ann. Rep. Alberta Hortic. Res. Centre* 1973, p. 60.

Mheen, H.J. van der. (1994) Comparison and development (during the ripening phase) of the seed- and carvone yield of (summer) caraway and dill. *Verslag – Proefstation voor de Akkerbouw en de Groenteteelt in de Vollegrond.*, No. **184**, 105 (in Dutch).

Mitchell, R.B. and Abernethy, R.J. (1993) Tolerance of clary sage, coriander and caraway to herbicides applied pre- and post-emergence. *Proc. 46th New Zealand Plant Protection Conf.*, Christchurch, New Zealand, 1993. pp. 24–29.

Munshi, A.M. (1991) *Carum carvi* L. – a new introduction from the Netherlands. *Crop Res. Hisar*, **4**, 309–311.

Munshi, A.M., Zargar, G.H., Baba, G.H. and Bhat, G.N. (1990) Effect of plant density and fertilized levels on the growth and seed yield of black zeera under rain-fed conditions. *Indian Cocoa, Arecanut and Spices J.*, **13**, 134–136.

Nordestgaard, A. (1986) Growing caraway (*Carum carvi* L.) for seed. Sowing and nitrogen rates. *Tidskr. Planteavl.*, **90**, 37–44 (in Danish).

Omar, M.O.M., Hussein, M.H., Darwish, Y.A. and Abdalla, M.A. (1991) Activity of flies and bees on flowering cumin, caraway and anise in relation to weather factors in Assiut and Sohag regions. *Assiut J. Agric. Sci.*, **22**, 81–92.

Ondrej, M. (1983) The occurrence of fungi on caraway (*Carum carvi* L.) in Czechoslovakia. Sb. UVTIZ (Ustav Vedeckotech. Inf. Zemed.) *Ochr. Rostl.*, **1**, 235–237 (in Czech).

Ondrej, M. (1988) New pathogen of caraway, dill, parsley and parsnip in Czechoslovakia – *Itersonilia pastinacae* Channon. *Bull., Vyzkumny a Slechtitelsky Ustav Zelinarsky Olomouc*, **32**, 109–112 (in Czech).

Pank, F., Marlow, H., Eichholz, E., Ennet, D. and Zygmunt, B. (1984) Chemical weed control in medicinal plants. Part 6. Caraway (*Carum carvi* L.). *Pharmazie*, **39**, 838–842 (in German).

Penka, M. (1978) Influence of irrigation on the contents of effective substances in officinal plants. *Acta Horticulturae*, No. **73**, 181–198.

Pinzaru, G. (1983) Chemical control of weeds in caraway (*Carum carvi* L.) and clary (*Salvia sclarea*). *Cercetari Agronomice in Moldava*, **2**, 81–83 (in Romanian).

Plescher, A. and Herold, M. (1983). The occurrence of diseases and pests of caraway (*Carum carvi* L.) in the years 1976 to 1981. *Nachrichtenbl. Pflanzenschutz* DDR. **37**, 12–18 (in German).

Prochazka, F. and Urbanova, J. (1972) Relations between quantitative characters of caraway. *Genetika a Slechteni*, **8**, 123–126 (in Czech).

Puschmann, G., Stephani, V. and Fritz, D. (1992) Studies on the variability of caraway (*Carum carvi* L.). *Gartenbauwissenschaft*, **57**, 275–277 (in German).

Putievsky, E. (1976) Yield components in *Carum carvi*. *Hassadeh*, **56**, 1702–1707 (in Hebrew).

Putievsky, E. (1977) Tests on the germination of caraway seeds. *Hassadeh*, **57**, 1413–1415 (in Hebrew).

Putievsky, E. (1979) Germination studies with seeds of seven species. *Aust. Seed Sci. Newsletter* 5: 147–149.

Putievsky, E. (1980) Germination studies with seeds of caraway, coriander and dill. Seed Sci. Technol., **3**, 245–254.

Putievsky, E. (1983) Effect of day-length and temperature on growth and yield components of three seed spices. *J. Hort. Sci.*, **58**, 271–275.

Putievsky, E. (1993) Selection and breeding of aromatic plants – old and new approaches. 1st World Congress on Medical and Aromatic Plants for Human Welfare, WOCMAP, Maastricht, The Netherlands, 1992. *Acta Horticulturae*, No. **330**, 137–141.

Putievsky, E. and Kuris, A. (1977) Developing system for growing annual *Carum carvi*. Hassadeh 57: 1776–1779 (in Hebrew).

Putievsky, E., Ravid, U., Dudai, N. and Katzir, I. (1994) A new cultivar of caraway (*Carum carvi* L.) and its essential oil. *J. Herbs, Spices and Medicinal Plants*, **2**, 81–84.

Putievsky, E., Ravid, U. and Sanderovich, D. (1984) Yield components, phenologica and aroma products of caraway. *Hassadeh*, **64**, 892–894 (in Hebrew).

Putievsky, E. and Sanderovich, D. (1985) Spacing and fertilization of caraway. *Hassadeh*, **65**, 1560–1561 (in Hebrew).

Regina, M. and Tulasi Raman. (1992) Biochemical changes in stored caraway seeds due to fungi. *Indian Phytopathol.*, **45**, 384.

Ricciardelli d'Albore, G.C. (1986) The pollinating insects of some Umbelliferae of agricultural and herbal interest (*Angelica archangelica, Carum carvi, Petroselinum crispum, Apium graveolens, Pimpinella anisum, Daucus carota, Foeniculum vulgare* v. *azoricum*). *Apidologie*, **17**, 107–124 (in French).

Rieder, J.B. and Reiner, L. (1972) High fertilizer rates for subalpine areas in combination with many-cut utilization and hot-air drying. *Bayer. Landwirtsch. Jahrb.*, **49**, 425–453 (in German).

Sarhan, A.Z. and El-Sayed, A.A. (1983) Effect of some growth regulators on growth and oil quantity of *Carum carvi* L. plants. *Ann. Agric. Sci* (Moshtohor), **19**, 411–417.

Sheidai, M., Ahmadian, P. and Poorseyedy, S. (1996) Cytological studies in Iran zira from three genus: *Bunium, Carum* and *Cuminum. Cytologia*, **61**, 19–25.

Smid, H.G. and Bouwmeester, H.J. (1993) Effect of light and pollination on the seed set and oil synthesis in caraway. *Trial results for 1992. Verslag Centrum voor Agrobiologisch Onderzoek-DLO*, 1993, No. **181**, 33 (in Dutch).

Toxopeus, H. and Bouwmeester, H.J. (1992) Improvement of caraway essential oil and carvone production in the Netherlands. *Industrial. Crops and Products*, **1**, 295–301.

Toxopeus, H. and Lubberts, H.J. (1994) Effect of genotype and environment on carvone yield and yield components of winter-caraway in the Netherlands. *Industrial Crops and Products*, **3**, 37–42.

Toxopeus, H., Lubberts, H.J., Neervoort, W., Folkers, W. and Huisjes, G. (1995) Breeding research and *in vitro* propagation to improve carvone production of caraway (*Carum carvi* L.). *Industrial Crops and Products*, **4**, 33–38.

Wander, J.G.N. and Zwanepol, S. (1994) Culture of caraway. *Teelthandleiding, Proefstation voor de Akkerbouw en de Groenteteelt in de Vollegrond.* No. **60**, 40 pp. (in Dutch).

Warakomska, Z. (1989) A rapid method for determining the presence of embryos in vegetable seeds of the family Umbelliferae. *Biul. Warzywniczy*, **33**, 59–64 (in Polish).

Weglarz, Z. (1983). Effect of agrotechnical factors on the transition of caraway (*Carum carvi* L.) from the vegetative to the reproductive phase. Part II. Effect of fertilization and soil moisture on the development and cropping of caraway. *Herba Polonica*, **29**, 21–26 (in Polish).

Wichtmann, E.M. and Stahl-Biskup, E. (1987) Composition of the essential oils from caraway herb and root. *Flav. Fragr. J.*, **2**, 83–89.

Zderkiewicz, T. (1971) Oil content at different stages of ripening in fruits of diploid and tetraploid cumin (*Carum carvi* L.). *Acta Agrobot.*, **14**, 121–127 (in Polish).

SECTION III
UTILIZATION

10. ALIMENTARY, CULINARY AND INDUSTRIAL USES OF CARAWAY

MICHAEL DACHLER

Bundesamt und Forschungszentrum für Landwirtschaft, Spargelfeldstrasse 191, 1220 Wien, Austria

10.1. CHEMICAL CONSTITUENTS AND TASTE

Spices provide of course only little contribution to the human nourishment. But they make food more tasty and enhances the appetite. The taste depends on certain chemical compounds mainly on essential oil content and composition. But not only the essential oil is responsible for the taste impression, also some of the other chemical components of caraway fruits are influencing the taste (Table 1).

Beside the mentioned chemical compounds, caraway contains tocopherol and tocotrienole, phenol-carbonic acids such as caffeic acid and gentisic acid, phenols and flavones (flavonole, quercetin etc.).

10.1.1. Description of the Taste

It is well known that the sense of taste can be realized as salty, bitter, sour and sweet with the tongue. Pungency is an impression of temperature and pain. Additionally we

Table 1 Main chemical constituents of biennial caraway fruits

Constituent	Content (in % of the fresh weight, 11–16% water)	
Protein	19–25	
Sugar (umbelliferose)	2–4	
Raw fiber	17–22	
N-free extractables	15–18	
Starch	4–5	
Fatty oil	8–30	thereof oleic 35–45
		linoleic 28–35
		petroselinic 25
Essential oil	3–7	thereof carvone 45–60
		limonene 35–55
		carveole > 1
Waxes, resins	1,5	
Sterine	0,5	thereof sito-, spina-sterine 48
		stigmasterine 33
Ash	4,5–7	

Table 2   Evaluation scale for the intensity of the three main taste
components (Triebe und Zobel 1991)

Taste intensity	Examples of spices for the taste		
	Spicy	*Pungent*	*Aromatic*
Overstrong	calmus	chillies	clove
Very strong	hop	peppers	nutmeg
Strong	pepper	pepper	pepper
Very distinct	lovage	mustard	*caraway*
Medium distinct	sage	onion	basilicum
Distinct	parsley	chives	dill
Very clear	dandelion	fennel	gooseberry
Clear	burnet saxifrage	peppermint	violet
Weak	portulak	Tripmadam	lime tree blossom
Very weak	marsh mallow	lemongrass	blue lungwort

are smelling with the olfactory nerve such odours as aromatic, fruity, flowery, resinous, but also foul or burning (Triebe and Zobel 1991).

The taste and odour of caraway fruits is due to their essential oil content, which consists of two main components, the D(+)-carvone (45–60%) and the D(+)-limonene (35–55%). The latter one is not so important for the taste.

To describe a taste is very difficult. Therefore many different definitions can be found. Some of them are given below:

Warm biting but pleasant, slightly minty, medicinal (Dzeziak 1989); tangy, tasty (Gerhardt 1990); the smell is peculiar aromatic, the taste is spicy (Melchior and Kastner 1974); aromatic, spicy and somewhat burning, the smell is strongly aromatic (Siewek 1990).

Herbs and spices are showing big differences concerning the ability to develop their aroma. But for the food industry it is very important to have standardized spices, which should not differ too much from one lot to the other. Seasonings are often composed from different spices. Triebe and Zobel (1991) proposed a system with regard on the intensity of the three main taste components (Table 2).

Corresponding to the intensity, the doses should be chosen (strong spices low amounts, weak spices high amounts). According to the authors, caraway should be on the average measured out from 0.7 to 1 g per portion or one great pinch between three fingers.

10.2. THE USE OF CARAWAY FRUITS AS SPICE

Caraway appears on the GRAS-list, a list of food additives which are generally recognized as safe by the FDA (Federal drug administration, USA) and is used as spice, seasoning, flavouring, oil and extract (McCaleb 1994).

10.2.1. Market Requirements

Caraway is mainly commercialized as a dried, fallen apart mericarp, as a whole, broken or ground. Whole fruits are 3–7 mm long, 1–2 mm thick and slightly curved. The surface is dark brown with five main yellow rips. Although the quality standards of commerce are not very high, light coloured and uniform fruits in size and shape are of advantage. The darker the fruits are, the lower the price. Supposedly caraway from the Netherlands is somehow larger than from Eastern Europe (Gerhardt 1990) and is called "straw caraway" (Seidemann 1993). The essential oil content is not as important as far the whole fruits are commercialized and the typical taste is obvious. In general, annual caraway has lower essential oil contents than biennial.

Good quality product consist of not more than 2% foreign bodies (Melchior and Kastner 1974), whereas Siewek (1990) pointed out, that good quality allows not more than only 0.2% foreign bodies.

Caraway powder has a yellowish-brownish colour and is not easy to distinguish from fennel powder. Under the microscope fragments of the endosperm there is often only the epidermal tissue to observe and the transverse cell layer is to realize whereas the secretory cavities are destroyed (Gassner et al. 1989).

Melchior and Kastner (1974) are reporting some adulterations, especially with gout weed (*Aegopodium podagraria* L.). The mericarp of gout weed is smaller, round and dark brown. Siewek (1990) but also Seidemann (1993) pointed out, that in the past some cases of adulteration took place, but nowadays not. But it may happen that lower quality products of caraway are mixed with better ones.

Caraway is sometimes mixed up with cumin (*Cuminum cyminum* L.). Both species look very similar, but their taste is distinctly different.

10.2.2. Microbial Contamination

Spices are sometimes heavily contaminated with different germs. Hartgen and Kahlau (1985) have examined different spices in household packages for their contamination with *Clostridium* and assessed the colony number on plate count agar. They found that some spices were particularly heavily contaminated, especially pepper, curry, paprika, basil, allspice, coriander and chillies. Caraway was not among the contaminated spices.

To prevent contamination, it is necessary that caraway is harvested when the fruits are dry enough to be stored. If they are not, an artificial drying is necessary. A disinfection is not usual.

A joint group of FAO/IAEA/WHO investigated chemical, sensorial and toxicological effects on irradiated spices. They found that an irradiation of caraway seeds with 10 kGy caused no toxicological problems, but the aroma was somehow altered (Schüttler and Bögl 1990).

10.3. PACKAGING

Spices, and as well as caraway fruits are normally packed in small glass bottles, coated paper bags or synthetic materials (PA/PE bags). The packing material should be impermeable for aroma compounds. Therefore the fruits have to be dry enough to prevent moulding (Stehle 1997). Spices should be stored on a dry, cool, and dark place in order to keep the aroma as long as possible.

10.4. THE USE OF CARAWAY IN THE KITCHEN AND FOOD INDUSTRY

10.4.1. The Use in the Kitchen

In Middle Europe caraway is a common spice in the kitchen but not everybody likes it. The use is limited to certain dishes and to certain regions (Gruber 1996).

Some examples are given below but the list must not necessarily be complete:

salad: cabbage, Chinese cabbage
vegetables (side dish): cabbage, potatoes, red cabbage, Savoy cabbage, sauerkraut
meat: pork (joint of pork), beef (pörkelt, goulash)
poultry: goose
cheese: fondue, gratin's, raclette

The reason why caraway is often used together with different species of cabbage can probably be seen in the anti-flatulent effect of caraway.

10.4.2. The Use in Seasonings

Most foods are not only seasoned with one spice but with a number of different tastes. The aroma industry is offering to the food industry a complete programme of different tastes produced of mixtures of spices, essential oils, oleoresins and all kinds off additives. The so called seasonings are compounds containing one or more spices or spice extracts, which, when added to the food, either during its manufacture or in its preparation, before it is served, enhance the natural flavour of the food and thereby increase its acceptance by the consumer. The following overlook is taken from Gerhardt (1990).

Bread seasonings: During the last years the seasoning of bread has increased considerably using anise, coriander, onion etc. and caraway. The amount added is about 10 g/kg. Higher amounts are linked with an unpleasant spice taste.

Chilli con carne: beside chillies, garlic, oregano, coriander also caraway can be found in this type of seasonings.

Curry: is a mixture of many different spices. Among others also caraway was found in Middle European curry mixtures.

Pickled vegetables mixtures: Especially for red cabbage-salad and sauerkraut-salad caraway is used beside many other spices.

Vegetable seasoning mixtures: cabbage, tomatoes, cucumbers, mushrooms, creamed potatoes

Grill seasonings: caraway additional to chillies, paprika, garlic etc.

Milk products: cheese and curd cheese.

Mustard: consists mainly of mustard meal, vinegar, salt, sugar but also from many spices, among them caraway.

Salad seasonings: the taste of the salad should be maintained or rather be enhanced by spices. In mixtures for meat, vegetable-, cheese-, potato-, and cabbage salads.

Soup seasonings: besides goulash- and potato soup caraway is not often used.

Sausage seasonings: for sausages many different spices are used. Caraway only in brawn (aspic). The amount added is about 5–6 g/kg sausage mass.

Table 3   Some of the known alcoholic beverages using caraway
(Clutton 1995)

Name	Origin	Remark
Akvavit	Scandinavia	Caraway with aniseed and fennel 40% alc.
Allash	Russia	Sweet kümmel with bitter almonds and aniseed
Cloc	Denmark	Kummel 31% alcohol, colourless
Kümmel	Netherlands	Caraway with some anise and cumin min 5% alcohol, one of the oldest liqueurs with digestive properties

In meat-, fish-, poultry-, and gamemeat-seasoning-mixtures caraway is normally not used.

In the anglosaxonian countries caraway is not a common ingredient in spice and seasoning preparations (Clark 1994).

10.4.3. The Use in Bakery Products

Caraway is not only mixed into white and ryebread but is also sprinkled on the dough before baking. Sometimes caraway fruits and essential oil is mixed together into the dough in order to get the aroma better spread out. The essential oil enhances the taste impression (Daffertshofer 1980).

10.4.4. The Use of Caraway in Alcoholic Beverages

The flavouring of different types of spirits and liquors has a long tradition. Caraway flavoured spirits are coming mainly from Denmark and other Scandinavian countries. Those products described as *akvavit* or *aquavit* are flavoured using neutral alcohol distillates of caraway and/or dill. Other flavourings, may be used, but the flavour of the drinks must be attributable to the distillates of caraway, aniseed or dill. Caraway is only added before distillation (Ney 1987). The use of essential oil is prohibited (Aylott 1995). Some well known alcoholic beverages which are on the market are listed in Table 3.

In American Gin the flavour additives always include juniper berries and usually cardamom as well as other botanicals such as caraway seeds and others (Cole and Nobel 1995).

10.5. THE ESSENTIAL OIL OF CARAWAY

The essential oil is obtained mainly of caraway fruits by steam or hydrodistillation. A good overlook about different distillation facilities is given by Lawrence (1995). The oil is colourless or slightly yellowish and toxic to humans even in relatively low amounts. Amounts of 4 g can already cause health disturbances for adults (Catizone *et al.* 1986). Carvone may also cause allergic effects (Roth and Daunderer 1996).

Instead of distillation of the ripe whole caraway fruit, it is also possible to chaff the umbels directly from the field into a container which is afterwards directly connected on a distillation unit. The essential oil yield gained in such a way was in the experiments of Hannig *et al.* (1988) about 48 l/ha. The early harvest had no negative influence on the oil composition (60% carvone in biannual caraway). Average essential oil yield (assessed on laboratory scale by Dachler *et al.* (1995) in variety tests) was around 70 kg/ha, with top yields of 160 kg/ha. In the contrary to earlier reported carvone contents (e.g. Lawrence 1992) they observed only 30–48% carvone in the tested biannual varieties.

Bazata *et al.* (1984) found, that during the distillation process the relation of carvone and limonene changed. At the beginning of the distillation the essential oil had higher amounts of carvone whereas at the end limonene predominated. The reason is that carvone as oxygen containing compound is several times more soluble in water than limonene.

The essential oil can be used in the food and cosmetics industries and has to a certain extent an antimicrobial and antioxidative effect.

10.5.1. Antibacterial Effects

Since long time human beings used herbs and spices for preventing food deterioration and pathogenic diseases. In general, plants from the *Liliaceae* family followed by *Myrtaceae*, *Cruciferae* and *Labiatae* showed the highest antimicrobial activity. While usually the antimicrobial effect of caraway is expected to be low, Morris *et al.* (1979) reports that together with essential oils of other botanicals the oil of caraway showed considerable effect.

Salzer (1982) examined six different spice extracts for their antibacterial properties. The effect of caraway oil was the lowest (pepper > nutmeg > maces > ginger > celery seeds > caraway). Caraway extract had some inhibitory effects against *Salmonella typhi murium* but even increased growth of *Bacillus cereus*. Against other germs caraway extract was rather indifferent (Salzer 1982).

Generally, gram-positive bacteria were more sensitive to the antimicrobial compounds in spices than Gram-negative ones, as the results of Farag *et al.* (1989) show. The oil of caraway exhibited moderate effect against bacteria, whereas the oils from clove and thyme were highly active.

10.5.2. Antifungal Effects of Caraway Oil

The constituents of caraway oil are showing only little effects against pests. It has been reported that merely limonene showed fungicide effects against *Lenzites saepiaria*, *Boletus variegatus* and *Trichoderma viride* (Degroot cit. by Lydon and Duke 1989).

10.5.3. Antioxidative Effects of the Oil and Oleoresins

Lipid peroxidation, a radical chain oxidation of unsaturated acids, causes various damage not only in living organisms but also in foods. Beside some chemicals it is well known that herbs and spices are proper materials to reduce or to inactivate rancidity in oils and fats. The antioxidative activity depends on the substrate of the oxidation, the preparation procedure and the oxidation test. As pointed out by Nakatani (1994), rosemary and sage as ground spices are remarkably effective, whereas caraway showed no real effectiveness.

Lee and Widmer (1994) compared the antioxidative capacity from some oleoresins such as rosemary, anise, dill and caraway as antioxidant for limonene oxidation. Rosemary oleoresins were most effective in inhibiting oxidation of limonene, whereas oleoresins from anise and caraway did exert a weak protective effect up to early stage of oxidation. Oleoresins from dill were nearly inactive against limonene oxidation.

10.5.4. Essential Oil in Food

Caraway oil is used in all major categories of foods including alcoholic and non-alcoholic beverages, frozen dairy desserts, candy, baked goods, gelatines and puddings, meat and meat products, condiments and relishes and others. Highest average maximum use level is reported to be about 0.02% in baked goods (Leung and Foster 1996).

10.5.5. Essential Oil Used in Cosmetics

Essential oil is also used as a fragrance component in cosmetic preparations including toothpaste, mouth wash, soaps, creams, lotions, and perfumes, with maximum use level of 0.4% reported in perfumes (Leung and Foster 1996).

10.5.6. Use of Essential Oil as Animal Food-additive

Vogt and Rauch (1991) evaluated the effect of essential oils as food additives for chicken. Neither with caraway oil nor with coriander, thyme, garlic or onion oils in different doses they found any significant influence on weight or feed efficiency. When checked at the end of the experiment in the organoleptic test, they did not found any typical taste or smell of the meat resembling to the corresponding essential oil in either of the cases.

It is described in elderly publications, that caraway plants may be used as animal food (pasture) for milking cows and sheep (Heeger 1956). Nowadays such a use is rather uncommon.

10.6. USE OF THE FATTY OIL FOR TECHNICAL PURPOSES

Petroselinic acid is an important raw material for oleochemical processes and can be easily cracked into lauric and adipinic-acid (Lechner 1997). As fatty oil from caraway seeds is comparatively rich in petroselinic acid and some origins also in fatty oil content, the fatty oil of caraway fruits may be also used in the oleochemical industry.

10.7. THE USE OF CARAWAY ROOTS AS VEGETABLE

Caraway roots can be cooked like carrots and have been used as a vegetable mostly in North European countries (Gistl and Nostitz 1932, Schuster 1992). The saccharose content is comparable to that of carrots. In Nordic countries precocious shooting is not desirable, if the roots with young leafs are used as vegetable in early spring (Weisaeth

1978). The frost resistant plant is rich in minerals and was in former days an early source of vitamin C. Nowadays still some, but very few people make caraway soup but it is not a common and even little known dish in Scandinavia. In Norway some efforts were made to reintroduce it as vegetable but up to now with little success (Dragland 1997).

10.8. USE OF THE BLOOMING PLANTS AS BEE-PASTURE

Like most Apiaceae (Umbelliferae), caraway is visited by bees rather frequently, and is also described in the literature as a good bee-pasture (Heeger 1956). However, no figures has been found about the real honey-yield capacity of caraway blossoms. As the acreage of caraway is normally not very large, it is not probably worthwhile for the beekeeper to produce "caraway honey" (despite the taste question).

REFERENCES

Aylott, R.I. (1995) Flavoured spirits. In Lea, A.G.H. and Pigott, J.R. (eds.) *Fermented beverage production*, Blackie academic & professional Glasgow.

Bazata, V., Plocek, J. and Suk, V. (1984) Contribution to the hydrodestillation of some essential oils of Apiaceae. *Intern. Vortragstagung Methoden und Verfahren der Züchtung, des Anbaus, der Sammlung und der industriellen Verarbeitung von Arznei und Gewürzpflanzen*. VEB Pharmazeutisches Werk Halle.

Catizone, P., Marotti, M., Toderi, G. and Tétényi, P. (1986) *Coltivazione delle piante medicinale e aromatiche*, Patron editore, Bologna.

Clark, M.W. (1994) Herbs and spices. In Underriner, E.W. and Hume, I.R. (eds.) *Handbook of industrial seasonings*, Blackie academic & professional London, Glasgow.

Clutton, D.W. (1995) Speciality products. In Lea, A.G.H. and Pigott, J.R. (eds.) *Fermented beverage production*, Blackie academic & professional Glasgow.

Cole, V.C. and Nobel, A.C. (1995) Flavor chemistry and assessment. In Lea, A.G.H. and Pigott, J.R. (eds.) *Fermented beverage production*, Blackie academic & professional Glasgow.

Dachler, M., Hackl, G., de Hueber, K. and Bailer, J. (1995) Einfluß von Sorte und Stickstoffdüngung auf Ertrag und Qualität von Kümmel. *Posterpresentation "Fachtagung Heil- und Gewürzpflanzen"* Freising-Weihenstephan.

Daffertshofer, G. (1980) Aromastoffe von Brot und Aromen für Feine Backwaren. *Gordian* **80**, 1–2, 17–20.

DeGroot, R.C. (1972) Growth of wood habitating fungi in the saturated atmospheres of monoterpenoids. Mycologia 64, 863–870. Cit. by Lydon, J. and Duke, S.O. (1989) The potential of pesticides from plants. In Craker, L.E. and Simon, J.E. (eds.) *Herbs, spices, and medicinal plants* Vol. 4, Phoenix.

Dragland, St. (1997) Apelsvoll Research Centre. Personal communication.

Dziezak, J.D. (1989) Spices. *Food technology*, 102–116.

Farag, R.S., Daw, Z.Y., Hewedi, F.M. and El-Baroty, G.S.A. (1989) Antimicrobial Activity of Some Egyptian Spice Essential Oils. *J. of Food Protection* **52**(9), 665–667.

Gassner, G., Hohmann, B. and Deutschmann, F. (1989) *Mikroskopische Untersuchung pflanzlicher Lebensmittel*. Gustav Fischer Verlag, Stuttgart.

Gerhardt, U. (1990) *Gewürze in der Lebensmittelindustrie*. Behr's Verlag, Hamburg.

Gistl, R. and Nostitz, A. von (1932) *Handelspflanzen Deutschlands, Österreichs und der Schweiz.* F. Enkeverlag, Stuttgart.

Gruber, P. (1996) *Das Gewürzkochbuch.* AT Verlag, Aarau.

Hahn, H. and Michaelsen, I. (1996) *Mikroskopische Diagnostik pflanzlicher Nahrungs-Genuß- und Futtermittel, einschließlich Gewürze.* Springer Verlag, Berlin, Heidelberg.

Hannig, H.-J., Herold, M. and Röhl, W. (1988) Resultate der Gewinnung etherischer Öle nach der Containertechnologie. *Drogenreport* **1**(1), 73–87.

Hartgen, H. and Kahlau, D.I. (1985) Bedeutung der Koloniezahl bei Haushaltsgewürzen. *Fleischwirtschaft* **65**(1), 99–102.

Heeger, E.F. (1956) *Handbuch des Arznei- und Gewürzpflanzenbaues.* Deutscher Bauernverlag, Berlin.

Lawrence, B.M. (1992) Progress in essential oils, Caraway oil. *Perfumer and Flavourist* **17**, 54.

Lawrence, B.M. (1995) The isolation of aromatic materials from natural plant products. In Tuley De Silva, K. (ed.) *A manual on the Essential Oil Industry,* Unido, Vienna.

Lechner, M. (1997) Industriegrundstoffe aus heimischen Ölpflanzen und die Perspektiven ihrer Nutzbarmachung 2. Forschungsprojektszwischenbericht Wien.

Lee, H.S. and Widmer, W.W. (1994) Evaluation of commercial oleoresins for inhibition of limonene oxidation. *Proc. Fla. state Hort. Soc.* **107**, 281–284.

Leung, A.Y. and Foster, S. (1996) *Encyclopedia of common natural ingredients used in food, drugs, and cosmetics.* John Wiley & Sons Inc., New York.

McCaleb, R.S. (1994) Food ingredient safety evaluation. In Charambolous, G. (Ed.) *Spices, Herbs and Edible Fungi.* Development in food science **34**. Elsevier Science B.V., Amsterdam.

Melchior, H. and Kastner, H. (1974) *Gewürze, Botanische und chemische Untersuchung.* Verlag Parey, Berlin und Hamburg.

Morris, J.A., Khettry, A. and Setz, E.W. (1979) Antimicrobial activity of aroma chemicals and essential oils. *J. Amer. Oil Chemists Soc.* **56**, 595–603.

Nakatani, N. (1994) Antioxidative and antimicrobial constituents of herbs and spices. In G. Charalambous (ed.) *Spices, Herbs and Edible Fungi. Developments in Food Science 34,* Elsevier Science B.V., Amsterdam, pp. 360.

Ney, K.H. (1987) *Lebensmittelaromen.* Behr's Verlag, Hamburg.

Roth, L. and Daunderer, M. (1996) *Giftliste.* 68. Erg. Lfg 8/96. Ecomed Verlag, Landsberg/Lech.

Salzer, U.-J. (1982) Antimikrobielle Wirkung einiger Gewürzextrakte und Würzmischungen. *Fleischwirtschaft* **62**(7), 885–887.

Schuster, W. (1992) *Ölpflanzen in Europa,* DLG-Verlag, Frankfurt a.M.

Schüttler, Ch. and Bögl, K.W. (1990) Chemische, sensorische und toxikologische Untersuchungen an bestrahlten Gewürzen. *Fleischwirtschaft* **70**(4), 431–440.

Siewek, F. (1990) *Exotische Gewürze.* Birkhäuser Verlag, Basel.

Seidemann, J. (1993) *Würzmittel-Lexikon.* Behr's Verlag. Hamburg.

Stehle, G. (1997) *Verpacken von Lebensmitteln.* Behr's Verlag, Hamburg.

Triebe, G. and Zobel, M. (1991) Bewertungsmodell von Würzmitteln als Grundlage für die sensorische Harmonisierung von Gewürzmischungen. *Drogenreport,* **6**, 15–23.

Vogt, H. and Rauch, H.-W. (1991) Der Einsatz einzelner ätherischer Öle im Geflügelmastfutter. *Landbauforschung Völkenrode* **41**(2), 94–97.

Weisaeth, G. (1978) Die Keimungskapazität der Samen von *Carum carvi* im Verhältnis zur Lebensdauer der Kümmelpflanze. *Seed Sci. & Technol.* **6**, 685–693.

11. PHARMACOLOGICAL USES AND TOXICOLOGY OF CARAWAY

ALA SADOWSKA and GRAZYNA OBIDOSKA

Department of Genetics, Plant Breeding and Biotechnology, Warsaw Agricultural University, Nowoursynowska 166, 02-766 Warsaw, Poland

11.1. HISTORY OF USING CARAWAY AS A REMEDY

The knowledge about usage of plants for medical purposes has always been one of the greatest treasures of any civilisation. The very important task for scientists working on medicinal plants is to preserve the primary med-botanical knowledge, commonly called folk medicine, before, as it was dramatically put by R.E. Schultes (1978), it gets buried together with the cultures who created it. Scientists from numerous countries work on gathering and organising the information widely spread in literature concerning the effect of various plants on human health. One of the plants is *Carum carvi*.

The history of using caraway as a remedy goes back to the days of the ancient Greeks, who probably knew and used the practical herb. It was mentioned by Plinius and Dioscorides, but today their correct identification of the species is uncertain.

In the Middle Ages *Carum carvi* was known and commonly used as a spice and a remedy. It was among the goods being a subject of trade in Europe. In 1410 Poland was undoubtedly one of the exporters of caraway, which is known from an old price list of spices passing through Gdańsk Harbour.

In the old days some medicinal plants were believed to posses extraordinary or even magical properties. That is why beside the quite usual uses, there were also some rather odd beliefs concerning various species. Caraway was one of them. Plants from *Umbelliferae* family in general were believed to provide protection against evil and witchcraft. In German tradition bread made with caraway fruits was used to drive away dwarfs and demons, and when brought into the house caused distress. The fruits were sometimes sprinkled with salt in coffins, as a protection against hexes, sorcery and demons. They were also considered effective in calming down restless children. A jar full of them placed under a cradle was said to have wonderful sedative powers (Mathias 1994). Moreover caraway, among other umbels, was one of the components of love potions, which made one fall in love with the person who had served it.

It was believed that for medical purposes caraway had to be collected on the 24th of June, when its healing powers were extremely high. The plant was used to treat a wide spectrum of diseases: indigestion, flatulence, eye disorders, coughs and even hysteria.

11.2. CARAWAY IN FOLK MEDICINE OF VARIOUS COUNTRIES

The range of *Carum carvi* is immense (from Northern Europe to the Mediterranean regions, Russia, Iran, Indonesia and North America). In numerous countries it is a very common species and, as a result, an integral part of their folk medicines. For example in Poland caraway is recommended as a remedy to cure indigestion, flatulence, lack of appetite, and as a galactagogue (Tyszyñska-Kownacka and Starek 1988). In Russia it is also used to treat pneumonia (Czikow and Laptiew 1982). In Great Britain and the USA it is regarded a stomachic and carminative. On The Malay Peninsula caraway is one of the nine herbs ground together and made into a decoction to be drunk at intervals after confinement, and in Indonesia the leaves mixed with garlic and spat on the skin are recommended to treat inflamed eczema (Perry 1980). Some of the properties are supported by scientific research and observations and are the reason for using caraway in contemporary medicine.

11.3. *CARUM CARVI* IN CONTEMPORARY MEDICINE

11.3.1. Fruit of Caraway as Medicinal Resource

Nowadays, as it used to be in the Past, *Carum carvi* is mainly known as a spice and the source of essential oil for the cosmetic industry. The role it plays in herbal medicine is a little less important, but also significant. The caraway fruit (*Fructus carvi*) is mentioned by pharmacopoeias of numerous European countries, USA and others. It is most of all used as a component of herbal mixtures recommended as a digestive, carminative and galactagogue.

Fruits of caraway as a herbal material should have following appearance: yellow to greyish-brown achenes, 3–6 mm long and ca. 1 mm thick, of a slightly crescent-shape with both ends pointed and 5 lighter colour ridges.

Caraway fruit should smell aromaticly and be quite spicy in taste. Normally it contains 12–20% oil, ca. 20% protein, ca. 20% carbohydrates, flavonoids, minerals and finally about 6% of the most important constituent, which is responsible for the medicinal and aromatic properties – the essential oil.

Oleum carvi is obtained from fruits by the means of distillation with water vapour. It is a transparent liquid, colourless or sometimes light yellow, with a pleasant smell and spicy taste (Rumiñska 1973). Its basic constituents are: carvone (50–65%), limonene (up to ca. 50%), dihydrocarveol, dihydrocarvone, sabinene, carveol, beta-pinene and other terpenes.

11.3.2. Effects

Fruits of caraway ingested orally produce an effect on the digestive tract, bile ducts, liver, and kidneys. They have spasmolytic properties towards the smooth muscles of the intestines, bile ducts and the sphincter regulating the flow of bile and pancreatic juices to the duodenum. They act as a cholagogue and increase the secretion of gastric juices, which results in appetite and digestion stimulation. *Fructus carvi* is a well known mild carminative helping to painlessly void the gas products of the metabolism. This is especially important for new-borns and infants. Caraway has also mild diuretic properties.

In women during lactation caraway fruit favours milk secretion. Although the component acting as a galactagogue hasn't been identified, it is present in the whole plant; leaves, stems and even roots. The milk secreted by women regularly drinking caraway tea has a beneficial anti-gripping effect on the digestive tract of breast-fed babies.

El Shobaki *et al.* (1990) reported advantageous effects of caraway extract on intestinal iron absorption.

The isolated essential oil of *Carum carvi* is administered orally as a component of mixed drugs. Pure is recommended for external use. It exhibits antifungal activity (Guerin and Reveillere 1985), kills dermal parasites and shows good inhibitory properties against *Staphylococcus aureus, Escherichia coli, Salmonella typhi, Vibrio cholerae* (Syed *et al.* 1987) and *Mycobacterium tuberculosis* (Mishenkova *et al.* 1985).

There is increasing evidence that a lot of plant derived substances may play an important role in cancer prevention. For example organosulphur compounds consumed in cruciferous vegetables (cabbage, cauliflower, broccoli) significantly decrease the incidence of cancer in humans (Oszmiański and Lamer-Zarawska 1996). Cancer chemopreventive agents produced by plants were firstly discovered during an international screening programme (1956–1981) started by The National Cancer Institute in Bethesda, USA. Its aim was to discover the natural compounds exhibiting anticancer properties (Sadowska 1991). Belman (1983) and Wattenberg and Sparnins (1989) reported the cancer chemopreventive activity of essential oils and their monoterpene components. The essential oils from caraway and dill (*Anethum graveolens* L.) as potentially anticarcinogenic substances were investigated by a group of scientists from Minneapolis: Zheng, Kenney and Lam (1992). Both oils exhibited high biological activity. The cancer chemopreventive property of caraway oil is probably due to the induction of the detoxifying enzyme glutathione S-transferase (GST). Zheng *et al.* (1992) reported that carvone and limonene are the compounds responsible for the above mentioned property while carvone exhibited even higher activity as a GST inducer. In 1993 Higashimoto *et al.* reported potent antimutagenic activity of caraway extracts against N-methyl-N'-nitro-N-nitrosoguanidine induced cancers in experimental animals.

11.3.3. Indications

Carum carvi is recommended as a remedy curing digestive tract disorders like: flatulence, eructation, stomach aches, constipation, lack of appetite, nausea. It is a mild drug, considered safe even for infants and the elderly. In small children caraway is used to treat flatulence and stomach aches, in the elderly for bile flow disorders, intestinal atony, vegetative neurosis (Ozarowski and Jaroniewski 1987).

Caraway fruit is recommended for breast-feeding women to enhance lactation and for the indirect, beneficial effect on the baby's digestive tract.

Externally pure essential oil or carvone may be used to treat dermal mycosis and scabies.

Caraway teas, and fruits added to food as a spice may also have a significant protective effect against anaemia (El Shobaki *et al.* 1990) and cancer (Zheng *et al.* 1992). The abundance of cancer chemopreventive substances in diet may even inhibit the early stages of carcinogenesis. Nature has equipped us with numerous "self-defence" mechanisms, which may be stimulated by certain natural substances. The knowledge of the mechanisms themselves and the biologically active compounds stimulating them is growing wider, and the control over the human "self-defence" system may be a question of the near future (Sadowska 1995).

11.3.4. Mode of Administration and Dose

Freshly pulverised caraway fruit is used for teas, infusions, and other galenical preparations like syrups and wines. *Fructus carvi* for internal use is often mixed with other herbs with similar or additional properties. Here are some recipes for caraway preparations according to Ozarowski and Jaroniewski (1987) and Lutomski and Alkiewicz (1993).

1. Caraway honey: mix ca. 1 g pulverised fruits with one tablespoonful of honey – take 2–4 times a day as a carminative.
2. Caraway tea: pour 1.5 glass (ca. 0.35 l) of boiling water over 1 tablespoonful of pulverised fruits in a thermos – drink ca. 0.5 glass 2–3 times a day, after meals.
3. Caraway syrup: pour 1 glass (ca. 0.25 l) of boiling water over 1 tablespoonful of pulverised fruits, leave covered for half an hour, then strain through a dense strainer or gauze, add honey – serve one teaspoonful after each meal as a carminative for children.
4. Carminative compositions:
   (a) Mix fruit of caraway (*Fructus carvi*), fruit of anise (*Fructus anisi*), pepper mint (*Folium menthae*), chamomile (*Anthodium chamomillae*) and thyme (*Herba thymi*) in equal proportions. Pour a glass (0.25 l) of boiling water over 1 tablespoonful of herbs, keep covered for half an hour, drink 0.5 glass, 2 times a day, after meals.
   (b) Mix fruit of caraway (*Fructus carvi*), fruit of anise (*Fructus anisi*), fruit of fennel (*Fructus foeniculi*) in equal proportions. Pour a glass (0.25 l) of boiling water over 1 tablespoonful of herbs, keep covered for half an hour, drink 0.5 glass, 2 times a day as a carminative and galactagogue.
   (c) Mix fruit of caraway (*Fructus carvi*), fruit of anise (*Fructus anisi*), fruit of fennel (*Fructus foeniculi*) and fruit of coriander (*Fructus coriandri*) in equal proportions. Pour a glass (0.25 l) of boiling water over 1 tablespoonful of herbs, keep covered for half an hour, drink 0.5 glass, 2 times a day.
5. Digestive preparations
   (a) Mix a double proportion of fruit of caraway (*Fructus carvi*) to fruit of fennel (*Fructus foeniculi*), herb of yarrow (*Herba millefolli*), herb of thistle (*Herba cnici*), root of liquorice (*Radix glycyrrhizae*) in equal proportions. Pour 2 glasses (0.5 l) of boiling water over 1.5 tablespoonful of herbs in a thermos, keep covered for one hour, drink 0.5 glass, about half an hour before meals.
   (b) Mix fruit of caraway (*Fructus carvi*), fruit of anise (*Fructus anisi*), pepper mint (*Folium menthae*), chamomile (*Anthodium chamomillae*) and thyme (*Herba thymi*) in

equal proportions. Pour 0.75 l of white, dry wine over 3 tablespoonfuls of herbs. Leave for two weeks, shaking from time to time – drink about 50 ml two times a day after meals.

6. Composition recommended for vegetative neurosis:
Mix fruit of caraway (*Fructus carvi*), flower of yarrow (*Flos millefolli*), root of valerian (*Radix valerianae*), herb of St. Johns Wort (*Herba hyperici*), leaves of Buckbean (*Folium menyanthidis*) and leaves of Bahu (*Folium melissae*). Pour 0.5 l of boiling water over 2 tablespoonfuls of herbs in a thermos and leave closed for half an hour, drink about 0.5 a glass, three times a day between meals.

7. Liniment for external use:
Dissolve 10 g of caraway essential oil (*Oleum carvi*) and 5 g of thyme essential oil (*Oleum thymi*) in 15 ml of 95% ethanol. Mix with 150 g of castor oil, or some other plant oil. Use for curing scabies and mycosis.

11.3.5. Ready-Made Compositions and Preparations

Caraway fruit is a component of a great number of galenical preparations and herbal compositions produced in many countries. Here are the examples of such medicines:

1. Laxative and carminative compositions:
Fito-Mix X (Poland), content: *Cortex frangulae, Rhadix rhei, Fructus carvi, Folium menthae.*
Normosan (Poland), content: *Fructus carvi, Folium menthae, Cortex frangulae, Folium sennae, Rhizoma agropyri, Fructus sambuci, Herba taraxacii cum Rhadici.*
Neonormosan (Poland), content: *Fructus carvi, Folium menthae, Radix rhei, Rhizoma agropyri, Fructus sambuci, Herba taraxacii cum Rhadici*
Purgaten (Poland), content: *Cortex frangulae, Fructus sambuci, Fructus carvi, Herba menthae, Folium sennae.*
Ortus species laxantes (Poland), content: *Cortex frangulae, Fructus sambuci, Fructus carvi, Herba menthae, Rhizoma agropyri, Anthodium chamomillae.*

2. Digestive and carminative compositions:
Digestosan (Poland), content: *Herba cnici, Folium menyanthidis, Radix althaeae, Folium menthae, Fructus carvi, Herba millefolli.*
Ortus species digestive (Poland), content: *Folium menyanthidis, Folium menthae, Fructus carvi, Herba millefolli, Anthodium chamomillae, Radix angelicae.*

3. Composition for acid peptic disorders:
Safoof-e-Satawar (India), content: *Glycyrrhiza glabra, Asparagus racemosus, Amomum subulatum, Carum carvi.*

4. Medicines containing caraway fruit extract:
Rhelax (Poland) – syrup: laxative, carminative, cholagogue.
Montana Haustropfen (Germany) – liquid: laxative, carminative, stimulates appetite, heals aches related to gall bladder disorders..
Schweden-Bitter (Germany) – liquid: laxative, carminative, gall bladder disorders, stimulates appetite.
Some other medicines containing *Carum carvi*: Tabulettae Laxantes, Gastrochol, Laxibel. Herbolaxine. Vitaflor carminative.

11.3.6. Caraway in Veterinary Medicine

Caraway is also used in veterinary medicine, but for animals the herb (*Herba carvi*) is a more popular remedy than the fruit. It contains a significant amount of essential oil and flavonoids. The decoction of the fruit and herb is used to cure gastrointestinal disorders like flatulence, indigestion, stomach aches and gripes. It promotes gastric secretion and stimulates appetite.

Caraway, like in humans, stimulates lactation in other mammals. Pulverised fruits or dry caraway herb are served mixed together with fodder to cows, mares in foal and other animals. Caraway as a galactagogue may also be of great importance for milk production. Voloshchuk *et al.* (1985) reported the significant increase of milk secretion in cows fed on pasture grounds enriched with caraway plants. The fat content in milk was lower in comparison with produced by cows fed on pastures without caraway.

The decoction from fruits or pulverised fruits are good remedies for rabbits, piglets and other animals against verminous disease. Gadzhiev and Eminov (1986) reported the effectiveness of caraway extract against trichostrongyle larvae in rams.

Caraway in veterinary medicine is also a remedy for external use. An ointment made from pulverised fruits mixed with vaseline is recommended against scabs, manges, mycosis and other dermal diseases. It may also be used to heal infected injuries or burns.

11.4. TOXICOLOGY OF CARAWAY

11.4.1. Toxicity of Caraway towards Man

The common opinion about the absolute safety of herbal drugs is undoubtedly wrong. It is obvious that biologically active compounds causing a therapeutic effect may be dangerous in certain situations and doses. As far as caraway is concerned, most authors agree that it shows no toxic effect towards people, and is well tolerated in medicinal doses and as a spice. However Lewis (1977) discussing the problem of allergy, mentioned carvone as sensitising substance, and classified *Carum carvi* among plants causing contact dermatitis.

The Umbelliferae in general is the family of plants rich in furocoumarins, which occur especially in fruits. Furocoumarins such as 5-methoxypsoralen (5-MOP) and 8-methoxypsoralen (8-MOP) are potent photosensitising, phototoxic, mutagenic and photocarcinogenic substances (Ceska *et al.* 1987). The consumption of some *Umbelliferae* fruits may be hazardous for health. It may result in severe dermatitis in the presence of sunlight. Fruits of *Carum carvi* have, among others, been tested for the content of 5-MOP and 8-MOP by Ceska *et al.* (1987). Using an ultrasensitive bioassay they detected just traces of these furocoumarins and concluded that the content of 5-methoxypsoralen and 8-methoxypsoralen in the fruits of *Carum carvi* is negligible.

Potentially hazardous for human health may be residues of nitrate, nitrite and pesticides in plants. Nitrate residues in herbs and vegetables can be transformed by bacteria to toxic nitrites which can cause blood circulation disorder and methemoglobinaemia. Generally nitrite and especially nitrate residues are present in herbs. Gajewska *et al.*

(1995) examined samples of 24 herbs and reported the content of $NaNO_2$ and KNO_3 in most of them. There were no exclusive data concerning fruits of *Carum carvi*, but fruits of fennel (*Fructus foeniculi*), although did not show the content of nitrites, did show relatively small but significant content of nitrates.

The analysis of pesticide residues in certain plants of medical importance (including *Carum carvi*) was performed by Dogheim *et al.* (1986). Gas liquid and thin layer chromatography tests showed that HCH was the main compound found in the tested samples. In the case of *Carum carvi* the residue did not exceed the maximum limit of 0.2 mg/kg.

The contamination of foods and herbs by fungi is a problem of great importance. This is mainly because of toxic and carcinogenic mycotoxins produced by numerous species. For example *Aspergillus flavus*, a very popular mildew fungus, produces aflatoxins which are the strongest known natural carcinogenic substances. They can be consumed with contaminated food and herbs, cumulated in liver and result in cancer. Pande and Bangale (1994) analysing the mycoflora associated with umbelliferous plants used in Ayurvedic medicines (*Foeniculum vulgare*, *Peucedanum graveolens* and *Carum carvi*) recorded a total of 50 species of fungi belonging to 30 genera. The most hazardous fungus being present on *Fructus carvi* was *Aspergillus* spp. responsible for aflatoxins and sterigmatocystin production. El Kady *et al.* (1995) using thin layer chromatography analysed 24 spices extracts. Aflatoxins and sterigmatocystin were discovered in some of them, including caraway. It means that caraway fruits and preparations made from fruits stored in favourable conditions for fungi, may be hazardous for human health.

As it was reported by Regina *et al.* (1992) fungi like *Aspergillus flavus*, *Aspergillus niger* and *Fusarium moniliforme* can cause biochemical changes in caraway fruits – reduce protein, carbohydrates and total oil, and increase fatty acids.

11.4.2. Toxicity of Caraway towards Bacteria, Fungi, Mites and Insects and its Advantages for People

Antibacterial and antifungal properties of caraway essential oil are used in therapy, but also in some other fields and may be of great importance for people.

Farag *et al.* (1989b) reported the inhibitory effect of six spice essential oils (including caraway) on 3 strains of Gram-negative bacteria and 4 Gram-positive bacteria. Gram-positive bacteria were more sensitive to the anti-microbial compounds in tested spices. This may be crucial for food preservation. Microbial spoilage of food is mostly inhibited by chemical food preservatives which are not fully safe for human health. Farag *et al.* (1989b) suggests that natural essential oils can be applied practically as anti-microbial agents which will prevent the deterioration of stored foods by bacteria and will not cause health problems to the consumer and handler.

The essential oils of several spices have strong antifungal properties and it is highly desirable to control mycotoxin formation during the storage of food. *Carum carvi* essential oil causes inhibition of mycelial growth and aflatoxin production of *Aspergillus parasiticus*. Chemicals such as potassium fluoride, acetic acid, potassium sulphite etc. also inhibit aflatoxin production but are hazardous for human health. Essential oils from spices are cheaper, natural, safe and widely used for flavouring. They can be applied as mould inhibitors to protect foods against aflatoxigenic fungi (Farag *et al.* 1989a).

Bang (1995) reported that several essential oils (including caraway oil) reduce significantly the infection of potato tubers by *Helminthosporium solani* and *Rhizoctonia solani*, which is important for storage.

Caraway is strongly toxic towards mites. Afifi and Hafez (1988) reported that petroleum ether, chloroform, acetone and methanol extracts of caraway fruits showed acaricidal properties towards *Tyrophagus putrescentiae*, which is a well known pest of stored products. The acetone extract was the most toxic one. In comparison with fenugreek (*Trigonella foenum-graecum*) and lupin (*Lupinus albus*) caraway showed higher toxicity towards *Tyrophagus putrescentiae*.

Domestic mites being one of the main causes of allergy are quite a significant problem. Ottoboni *et al.* (1992) tested the toxicity of 10 essential oils from different plants towards the most popular house mites *Dermatophagoides pteronyssinus*, *D. farinae*, *Euroglyphus maynei*, *Acarus siro*, *Tyrophagus putrescentiae*, *Glycyphagus domesticus*, *Lepidoglyphus destructor* and *Gohiera fusca*. From all 10 volatile oils tested, the caraway oil was among the 4 most effective ones. The authors conclude that using essential oils combined with cleaning agents could help reduce the cause of allergy in people. The strong toxicity towards mites was also proved by Watanabe *et al.* (1989) who tested 52 essential oils. Caraway was among the 6 most powerful ones. Isolated d-carvone showed very high activity thus being responsible for acaricidal property of caraway essential oil to a great extend.

Caraway is also toxic for some insects. Petroleum ether extract of caraway causes inhibition of larvae development in *Musca domestica*, *Culex pipiens fatigans* and mosquito (Deshmukh *et al.* 1987). The caraway extract has also the feeding deterrent activity in 5th-instar larvae of *Spodoptera littoralis* (Antonious and Hegazy 1987). These facts offer the hope for new, non-toxic to people botanical insecticides against the most popular insects and crop pests.

REFERENCES

Afifi, F.A. and Hafez, S.M. (1988) Effect of different plant extracts on the toxicity and behaviour of *Tyrophagus putrescentiae* Schrank (Acari: Acaridae). *Annals of Agricultural Science Cairo*, **33**(2), 1375–1385.

Antonius, A.B. and Hegazy, G. (1987) Feeding deterrent activities of certain botanical extracts against the cotton leafworm, *Spodoptera littoralis* (Boisd.). *Annals of Agricultural Science, Ain Shams University*, **32**(1), 719–729.

Bang, U. (1995) Natural plant extracts – control of fungal pathogens of potato. *Svenska Vaxtskyddskonferensen: Jordbruk – Skadedjur, vaxtsjukdomar och ogras*, **36**, 371–381.

Belman, S. (1983) Onion and garlic oils inhibit tumor promotion. *Carcinogenesis*, **4**, 1063–1065.

Ceska, O., Chaudhary, S.K., Warrington, P.J. and Ashwood-Smith, M.J. (1987) Photoactive furocoumarins in fruits of some umbellifers. *Phytochemistry*, **26**(1), 165– 169.

Czikow, P. and Laptiew, J. (1982) *Rośliny lecznicze i bogate w witaminy*, PWRiL, Warszawa.

Deshmukh, P.B. and Renapurkar, D.M. (1987) Insect growth regulatory activity of some indigenous plant extracts. *Insect Science and its Application*, **8**(1), 81–83.

Dogheim, S.M., Almaz, M.M., Takla, N.S. and Youssef, R.A. (1986) Multiple analysis of pesticide residues in certain plants of medical importance. *Bulletin of the Entomological Society of Egypt, Economic*, 15, 157–163.

El Kady, I.A., El-Maraghy, S.S.M. and Mostafa, M.E. (1995) Natural occurrence of mycotoxins in different spices in Egypt. *Folia Microbiologica*, 40(3), 297–300.

El Shobaki, F.A., Saleh, Z.A. and Saleh, N. (1990) The effect of some beverages extracts on intestinal iron absorption. *Zeitschrift fur Ernahrungswissenschaft*, 29(4), 264–269.

Farag, R.S., Daw, Z.Y. and Abo-Raya, S.H. (1989a) Influence of some spice essential oils on *Aspergillus parasiticus* growth and production of aflatoxins in a synthetic medium. *Journal of Food Science*, 54(1), 74–76.

Farag, R.S., Daw, Z.Y., Hewedi, F.M. and El Baroty, G.S.A. (1989b) Antimicrobial activity of some Egyptian spice essential oils. *Journal of Food Protection*, 52(9), 665–667.

Gadzhiev, Y.G. and Eminov, R.Sh. (1986) Action of medicinal plants on gastrointestinal nematodes of sheep. *Byulleten Vsesoyuznogo Instituta Gelmintologii im. K.I. Skryabina*, 44, 12–16.

Gajewska, R., Nabrzyski, M. and Wierzchowska-Renke, K. (1995) Zawartosc azotynów i azotanów w ziolach. *Wiadomosci Zielarskie*, 6, 13–14.

Guerin, J.C. and Reveillere, H.P. (1985) Antifungal activity of plant extracts used in therapy. II Study of 40 plant extracts against 9 fungal species. *Annales Pharmaceutiques Francaises*, 43(1), 77–81.

Higashimoto, M., Purintrapiban, J., Kataoka, K., Kinouchi, T., Vinitketkumnuen, U., Akimoto, S., Matsumoto, H. and Ohnishi, Y. (1993) Mutagenicity and antimutagenicity of extracts of three spices and a medicinal plant in Thailand. *Mutation Research*, 303, 135–142.

Lewis, W.H. (1977) *Medical botany. Plants affecting man's health*, Wiley Interscience, New York London Sydney Toronto.

Lutomski, J. and Alkiewicz, J. (1993) *Leki roœlinne w profilaktyce i terapii*, PZWL, Warszawa.

Mathias, M.E. (1994) Magic, myth and medicine. *Economic Botany*, 48(1), 3–7.

Mishenkova, E.L., Petrenko, G.T. and Klimenko, M.T. (1985) Effect of antibiotic substances from higher plants on mycobacteria. *Mikrobiologicheskii Zhurnal*, 47(1), 77–80.

Oszmiañski, J. and Lamer-Zarawska, E. (1996) Substancje naturalne w profilaktyce chorób nowotworowych. *Wiadomosci Zielarskie*, 7/8, 9–11.

Ottoboni, F., Rigamonti, I.E. and Lozzia, G.C. (1992) House mites prevention in Italy. *Bollettino di Zoologia Agraria e di Bachicoltura*, 24(2), 113–120.

Ozarowski, A. and Jaroniewski, W. (1987) *Rosliny lecznicze i ich praktyczne zastosowanie*, Instytut Wydawniczy Zwiazków Zawodowych, Warszawa.

Pande, A. and Bangale, S. (1994) Mycoflora associated with umbelliferous plant parts used in Ayurvedic medicines. *Journal of Economic and Taxonomic Botany*, 18(1), 83–86.

Perry, L.M. (1980) *Medicinal Plants of East and Southeast Asia*, The MIT Press, Cambridge, Massachusetts and London, England.

Regina, M. and Tulasi-Raman, Raman, T. (1992) Biochemical changes in stored caraway seeds due to fungi. *Indian Phytopathology*, 45(3), 384.

Rumiñska, A. (1973) *Rosliny lecznicze. Podstawy biologii i agrotechniki*, PWN, Warszawa.

Sadowska, A. (1991) *Rosliny i roslinne substancje przeciwnowotworowe*, Pañstwowe Wydawnictwo Naukowe, Warszawa.

Sadowska, A. (1995) *Rosliny i roslinne substancje rakotwórcze*, Fundacja "Rozwój SGGW", Warszawa.

Schultes, R.E. (1978) *Atlas de plantes hallucinogenes du mond*, Ed. de Aurora, Montreal.

Syed, M., Khalid, M.R., Chaudhary, F.M. and Bhatty, M.K. (1987) Antimicrobial activity of the essential oils of the *Umbelliferae* family. Part V. *Carum carvi, Petroselinum crispum* and *Dorema ammoniacum* oils. *Pak. J. Sci. Ind. Res.*, 30(2), 106–110.

Tyszyñska-Kownacka, D. and Starek, T. (1988) *Ziola w polskim domu*, Wydawnictwo Watra, Warszawa.

Voloshchuk, N.M., Riznichuk, S.T., Krul, M.I. and Marchuk, M.T. (1985) Sward type, pasture yield and milk production. *Kormoproizvodstvo*, 9, 34–35.

Watanabe, F., Tadaki, S., Takaoka, M., Ishino, M. and Morimoto, I. (1989) Killing activities of the volatiles emitted from essential oils for *Dermatophagoides pteronyssinus*, *Dermatophagoides farinae and Tyrophagus putrescentiae*. *Shoyakugaku Zasshi – Japanese Journal of Pharmacognosy*, **43**(2), 163–168.

Wattenberg, L.W., Sparnins, V.L. and Barany, G. (1989) Inhibiton of N-nitroso-diethyloamine carcinogenesis in mice by naturally occurring organosulphur compounds and monoterpenes. *Cancer Research*, **49**, 2689.

Zheng, G.-q., Kenney, P.M. and Lam, L.K.T. (1992) Anethofuran, carvone, and limonene: Potential cancer chemopreventive agents from dill weed oil and caraway oil. *Planta Medica*, **58**(4), 338–341.

12. APPLICATION OF S-CARVONE AS A POTATO SPROUT SUPPRESSANT AND CONTROL AGENT OF FUNGAL STORAGE DISEASES

KLAASJE J. HARTMANS, KOOS OOSTERHAVEN,
LEON G.M. GORRIS and EDDY J. SMID

*Agrotechnological Research Institute (ATO-DLO), P.O. Box 17,
6700 AA Wageningen, The Netherlands*

12.1. INTRODUCTION

Caraway (*Carum carvi*) seed has a long tradition of being used as a spice to improve the flavour and fragrance quality of food. There is also a long tradition of commercial distillation of the essential oil from caraway seeds. The main components of caraway seed essential oil are S-carvone and R-limonene and the oil is mainly used in oral preparations to overcome unpleasant odour or taste. The aromatic profile of S-carvone is described as fruity, sweet, musty, earthy and accounts for the "spearmint-like" description in the profile of caraway (Heath 1973). In addition to the aromatic characteristics, S-carvone is known to possess antifungal (Janssen 1989, Smid *et al.* 1994), insect repellent (Zuelsdorff and Burkholder 1978) and plantgrowth inhibiting activities (Bournot 1949). These features have encouraged researchers to develop applications of S-carvone for crop protection and plant growth regulation.

12.1.1. Crop Protection

Protection of crop and ornamental plants from noxious insects, nematodes, mites, microbial pathogens and weeds is indispensable to modern agriculture. Despite intensive control efforts, about 50% of the worlds crops are lost to these organisms, at an estimated cost of about 400 billion dollars (Ausher 1996). Ever since the advent of synthetic pesticides in the 1940s, modern crop protection has been largely based on chemical control. Mounting environmental concerns and pest control failures have made it increasingly clear that the use of toxic pesticides in agriculture should be drastically reduced all over the world.

Natural methods of crop protection can contribute to an appropriate pest management, the more as they are minimizing the risks for farmer, consumer, and the environment. Furthermore, scientific investigation can do much to improve their applicability and effectiveness, thus modernizing their use. In that aspect there is an increasing interest for plant essential oil components with for instance antifungal, insecticidal or growth inhibiting activities.

Plant essential oils are released from the plant to its environment by volatilization, and consequently some of these components are being able to interfere with other

175

plants. This phenomenon is called allelopathy. The term allelopathy was coined by Molisch in 1937 to refer to biochemical interactions between all types of plants, including both inhibitory and stimulatory interactions. Allelopathy is a well known fact in ecosystems and specially in desert ecosystems (Friedman 1987). Plants with allelopathic potential are often adult perennials, members of the Compositae or Labiatae, that are capable of reducing germination and/or growth of various annuals or of their own seedlings (autotoxicity). Variable ecological conditions may trigger allelopathy in one habitat and abrogate it in another. Volatile compounds accumulate in the soil during the long dry summer and in the winter, when germination commences, they will be released into the soil microsphere, where they inhibit growth of other plants.

Allelochemicals have already been shown to impose numerous impacts in cultivated and natural ecosystems. Many types of chemicals are involved in allelopathy, including essential oils, which contain a great variation of compounds most of them belonging to terpenes or terpenoids.

Plant essential oils can have, besides allelochemical properties, many other biological functions such as attracting or repelling insects, antimicrobial activity or making particular plants or part of plants unattractive as food for animals.

S-carvone from the essential oil of caraway seed has recently been developed as a crop protection product on potatoes, because of its interesting allelochemical properties as a sprout growth suppressant and antifungal agent on pathogens causing storage diseases.

12.2. POTATO STORAGE

12.2.1. Storage Conditions

The potato is the world's fourth most important food crop after the leading three rice, maize and wheat. The demand for potatoes is relative stable in Western Europe and the USA. In Eastern Europe we can expect dramatic increases in productivity per unit area if current know-how and technological expertise are applied. Furthermore, an increasing proportion of this global crop is being produced in the less-developed countries, now approximately 1/3 of the total, and rising. The proportion of the crop utilized in processed products like French fries, chips (crisps) etc. is worldwide increasing.

A potato tuber is in fact a modified stem with a shortened (and broadened) axis. The apex and the eyes can be distinguished on the tubers. The eyes often contain not only a main bud but also a few small lateral buds (Figure 1). As many as 20 or more buds may be present on a tuber. Control of spout growth from these buds is very important when tubers have to be stored for a long time. Potato storage is essential in all parts of the world where the crop can be grown for only parts of the year. After storage the potatoes should be firm and unwilted. Potato tubers cease to be firm when evaporative loss from an initially turgit tuber reaches about 5%. Moisture losses are much greater in potatoes that have sprouted. Apart from direct weight loss in the form of sprouts, Burton (1955) calculated that extra weight losses occurred due to the fact that the epidermis of the sprout is about 100 times more permeable to water as the surface of the tuber.

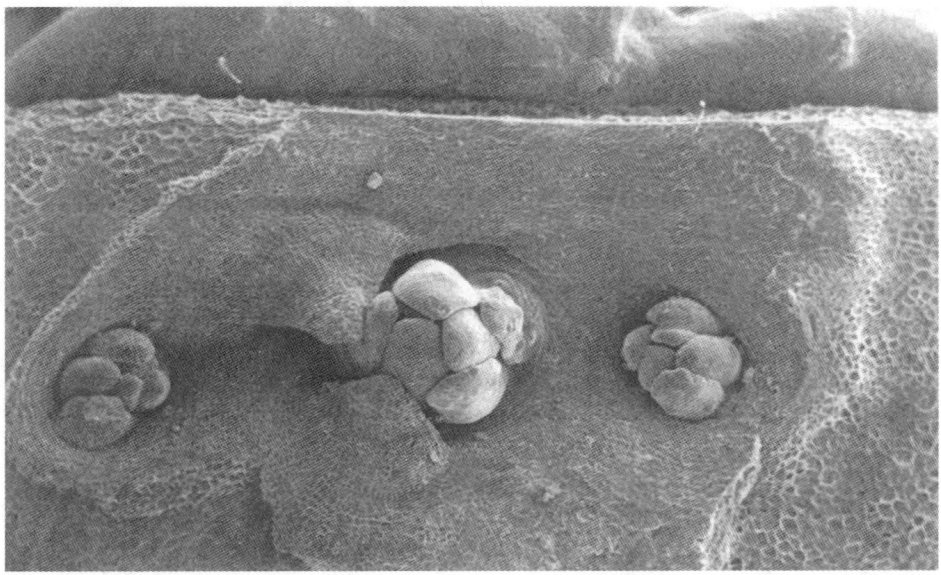

Figure 1 Scanning electron micrograph of a single potato eye, showing the original axillary bud of the leaf scale and two second order axillary buds of its two leaf primordia. (By courtesy of Chris van der Schoot, ATO-DLO)

Storage at low temperature can control sprouting for a long time, but causes a marked increase in reducing sugars, particularly when stored at temperatures below 6°C. Below 10°C reducing sugars are formed to an increasing degree (Burton 1965a). The reducing sugar content is very important where potatoes are processed into fried products. On frying, the potato products darken owing to the reaction between the reducing aldehyde groups of the sugars glucose and fructose and the amino groups of the free amino acids: this is known as the Maillard reaction. This darkening may be so intense as to greatly reduce the value of the end product. Of all the factors that determine the amount of reducing sugars during storage and hence the suitability for processing, cultivar and storage temperature are probably the most important.

In practice storage temperatures are generally maintained at 6–8°C. To keep the temperature of the potatoes during storage at 6–8°C, regular cooling is needed, because the respiration of the tubers during storage results in a permanent uptake of oxygen and a release of carbon dioxide, water and energy in the form of heat. Cooling with outside air is an efficient and relative inexpensive method, which is often used in potato storage. Outside air can be used to cool the potatoes only during periods in which the ambient temperature regularly falls below the required storage temperature. Consequently during such, sometimes short, periods high ventilation rates are used to be able to cool down the potatoes as quick as possible (Rastovski 1987).

In order to be able to control sprout growth during longterm storage at these temperatures, chemical sprout growth suppressants are widely used. In general, the high ventilation rates of the cooling systems make regular reapplication of sprout suppressants during storage necessary.

12.2.2. Chemical Sprout Suppressants

The principle sprout inhibitor used worldwide today is isopropyl N-(3-chlorophenyl)-carbamate (Chlorpropham; CIPC) (Marth and Schultz 1952). CIPC was originally developed as a herbicide, but proved to have good potato sprout suppressing properties. Of minor importance is isopropyl N-phenylcarbamate (Propham; IPC). Sometimes mixtures of CIPC and IPC are used ((C)IPC). The effectiveness of (C)IPC in preventing sprouting varies and occasionally the inhibitors induce undesirable side effects. For example, both carbamates are very persistent chemicals in storage buildings. Considerable residues can built up in the fabric of the stores (Boyd and Duncan 1986) and is a source of headspace CIPC within the stores. When used for seed potato storage this may affect the viability of the tubers.

12.2.3. Natural Sprout Inhibitors

Farmers of the high Andes near Cuzco and on the Altiplano near Puno use the twigs of the muña plant to protect potatoes in store (Ormachea 1979). Muña is a perennial, shrublike plant and includes the genera *Minthostachys* and *Satureja*. Of the *Minthostachys* twelve species are known from Venezuela to Argentina. The plant contains a high amount of essential oil and is effective as insect repellent, larvicidal agent and inhibitor of potato tuber sprouting in store (Stoll 1986) and for these purposes used ever since the precolumbian period. The volatile components of muña essential oil are emanating from the leaves into the surrounding air. The major components of the muña essential oil are pulegone (45%) and menthone (18%), (Aliaga and Feldheim 1985, Baerheim Svendsen *et al.* 1987). The sprout suppressing effect of muña oil is primarily caused by pulegone (Hartmans, unpublished results). Potato sprout suppression by natural volatile organic compounds including pulegone and carvone have also been reported by Meigh (1969), Beveridge *et al.* (1981), Hartmans and Van Es (1988) and Vaughn and Spencer (1991, 1993).

Volatile substances are also released from the potato tubers during storage and if not eliminated by ventilation, may totally inhibit sprouting (Burton 1952, 1965b; Burton and Meigh 1971). Meigh *et al.* (1973) identified some of the aromatic compounds evolved by stored potatoes and determined benzothiazole, 1,4-dimethylnaphthalene and 1,6-dimethylnaphthalene, all of which are comparable potent inhibitors of sprout growth in the potato tuber.

Many natural occurring monoterpenenes and other volatile compounds possess a good sprout suppressant ability. We have investigated the sprout inhibiting properties of a number of monoterpenes in small scale experiments (Hartmans unpublished). A few monoterpenes, including S-carvone, were found to suppress sprout growth under these conditions for more than a year, depending on the amount applied. Besides good sprout suppressing properties, S-carvone showed no negative effects on potato quality. Furthermore, S-carvone is placed on the list of safe food ingredients (GRAS-Generally Recognized As Safe list) by the Flavoring Extract Manufacture Association in 1965 and was approved by the Food and Drug Association for use in food in 1961 (Chan 1990 and references therein). For these reasons and aspects as agronomic perspectives, S-carvone was selected for further research to be developed as a natural sprout suppressant of potatoes.

In 1989 a research programme was started in The Netherlands to develop the practical application of S-carvone as a sprout suppressant of potatoes (Hartmans *et al.* 1995).

Since 1994, S-carvone is registered in The Netherlands as a commercial sprout suppressant for ware potatoes under the tradename 'Talent' (95% S-carvone) and delivered by LUXAN B.V., The Netherlands.

12.3. S-CARVONE AS A SPROUT SUPPRESSANT FOR WARE POTATOES

S-carvone acts in the vapour phase to prevent sprouting. S-carvone has a high vapour pressure. During and after application it evaporates relatively quickly. Losses of S-carvone in the vapour phase are caused by: (a) regular ouside air ventilation; (b) adsorption at potatoes, soil, building materials etc.; (c) bioconversion by potatoes and microorganism. S-carvone ('Talent') therefore has to be reapplied at regular intervals during storage.

Application of the liquid 'Talent' can be done with different techniques, e.g. fogging equipment (Swingfog, Pulsefog) or a nozzle system, by which the 'Talent' is brought into the air-stream used for cooling. During and after application the ventilation system is used as internal ventilation system to distribute the 'Talent' well. External ventilation is excluded for about 2 days after application to avoid exceptional losses.

12.3.1. Sprout Suppression of Ware Potatoes during Long Term Storage

The effects of 'Talent' were compared with standard treatments of (C)IPC-containing products, applied according to the label $(20\,g \cdot 1000\,kg^{-1} \cdot season^{-1})$. 'Talent' and (C)IPC products were in general applied with a fogging apparatus (Swingfog).

From small scale laboratory experiments it was known that the S-carvone vapour acts as sprout suppressant. Consequently, the S-carvone concentration of the storage atmosphere should be kept to a certain level in order to prevent sprouting. To be able to check the S-carvone concentrations of the stores, air samples of the stores were analysed by Gas Chromatography (Hartmans *et al.* 1995). Periodical 'Talent' treatments $(100\,ml \cdot 1000\,kg^{-1})$ in semi-practical bulk stores of 14 tonnes showed high S-carvone concentrations just after application, reaching in general the temperature dependent saturation limit. Thereafter the S-carvone concentration decreased within a few weeks to a low level (Figure 2). The average S-carvone concentration between two interval treatments depended on the amount of 'Talent' applied and was on average somewhat higher in the middle of the storage season when there was less external ventilation due to lower outside air temperatures (winterperiod).

S-carvone is able to inhibit the sprout growth of potatoes very effectively (Figure 3). When applied at regular intervals, 'Talent' was able to suppress sprout growth during prolonged storage with the use of outside air cooling (Table 1). Compared to standard treatments with (C)IPC containing products, 'Talent' had comparable or even better

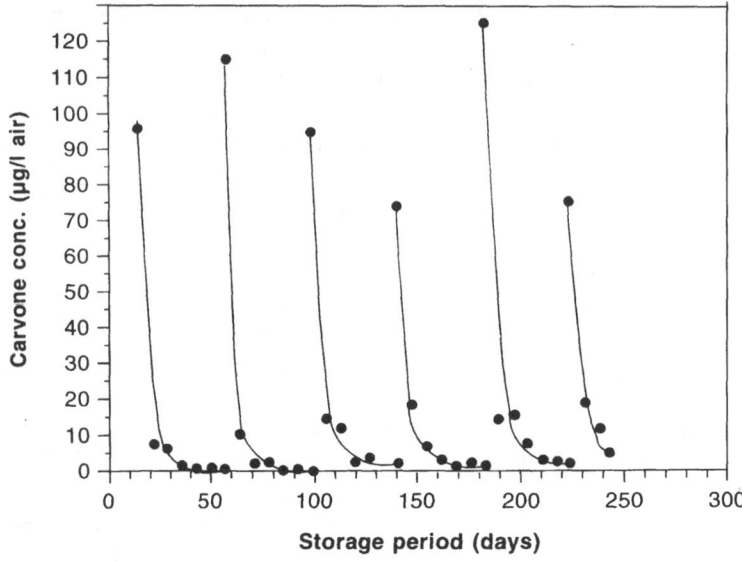

Figure 2 Influence of repeated S-carvone applications $(100 \, \text{ml} \cdot 1000 \, \text{kg}^{-1} \cdot 6 \, \text{weeks}^{-1})$ on the S-carvone concentration in the storage atmosphere

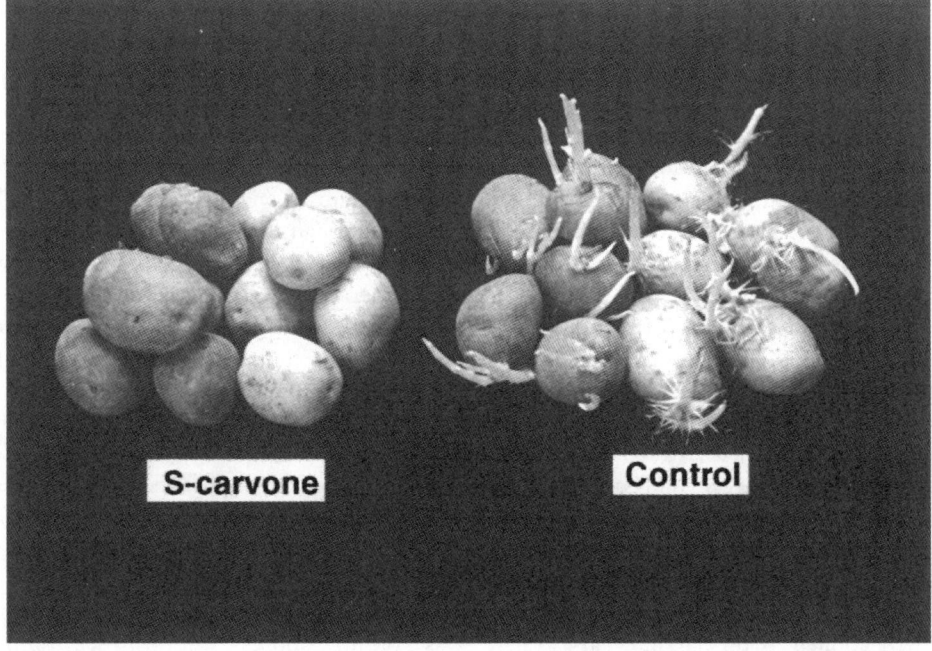

Figure 3 Sprout suppression of two potato cultivars (Bintje and Désirée) with S-carvone (See Color Plate II)

Table 1 Sprout suppression effects of 'Talent' (95% S-carvone) and
(C)IPC (30%) at unloading time. Data represent average values obtained
from several storage experiments carried out from 1990 to 1996.
The main cultivar used, amongst others, was Bintje

Sprout suppressant	Dosage per application ml · $1000 \, kg^{-1}$	Application intervals	Total amount $1000 \, kg^{-1}$	Average storage period (days)	Sprout suppression (*)
(C)IPC	20 ml	variable	20 g	240	4.5
Talent	100 ml	6 weeks	570 ml	260	6
Talent	50 ml	3 weeks	550 ml	260	6
Talent	75 ml	6 weeks	430 ml	260	3
Talent	100 ml	9 weeks	380 ml	260	4
Talent	10 ml	1 week	350 ml	260	2.5

(*)The mean sprout suppression effect was determined and expressed on a sliding
scale with the following indications: (1) bad, (2) insufficient, (3) almost sufficient, (4)
sufficient, (5) amply sufficient, (6) good and (7) excellent.

sprout inhibiting properties. During several storage seasons the best results were
achieved when 100 ml 'Talent'·$1000 \, kg^{-1}$ was applied every 6 weeks (6 times in total).
When treated this way, potatoes could be stored for a long period with good sprout suppression effects. Using other 'Talent' treatments resulted in reduced sprout suppression.

12.3.2. Quality Aspects of S-carvone Treated Ware Potatoes During Storage

The main quality aspects of S-carvone treated 'ware potatoes' during storage are: the
S-carvone residue levels, processing quality and sensory quality of cooked or processed
potatoes.

12.3.2.1. S-carvone Residue Levels

Most S-carvone was found in the peelings (on average 90%) of the tubers and only a
small amount in the peeled tubers. S-carvone is a rather lipophilic compound and will
be easily adsorbed at the suberized potato periderm, whereof it can be removed with
outside air ventilation. At the end of the storage period the raise in outside air temperature causes an increase in the total amount of ventilation hours if cool air is available.
This results in a decrease of S-carvone residue levels. At unloading time the average
S-carvone concentration in the whole, washed but unpeeled, tubers ranges in most
cases between 1 and 2 mg·kg^{-1} fresh weight and never exceeds the level of 5 mg·kg^{-1}
fresh weight. The S-carvone residues vary during storage (Figure 4). The S-carvone
residues depend on the: (a) storage time; on average higher in the middle of the storage
season when there was less external ventilation (winterperiod) (Figure 4), (b) cultivar
properties like tuber size; small tubers have higher residue values then big ones when
utilizing the same amount of S-carvone, (c) dosage applied, (d) time between application and sampling, (e) hours of ventilation with outside air after application, (f) storage

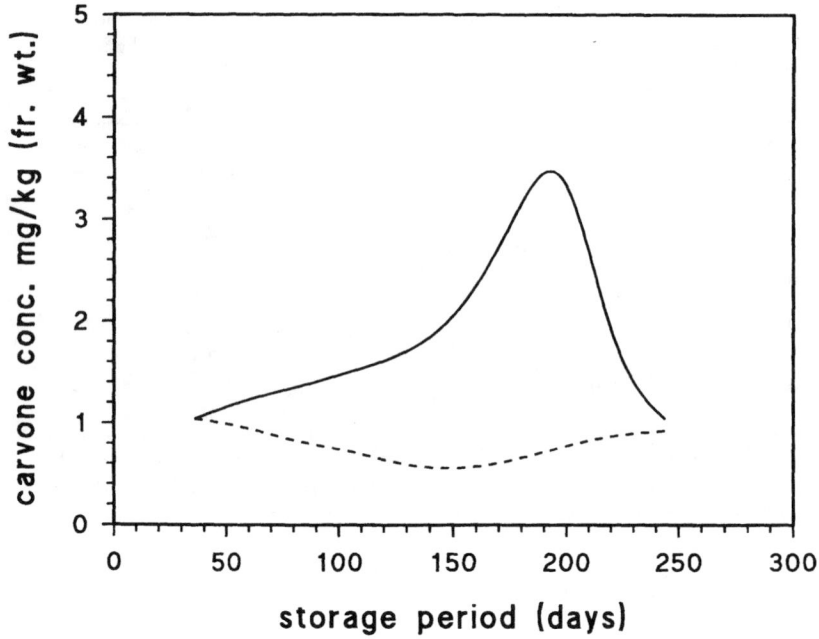

Figure 4 Influence of repeated S-carvone applications $(100\,\text{ml}\cdot 1000\,\text{kg}^{-1}\cdot 6\ \text{weeks}^{-1})$ on the S-carvone residue of tubers stored at 5–7°C (solid line) or 7–10°C (dotted line)

temperature; faster metabolization at higher average storage temperature (Figure 4), (g) location of the tubers in store; in general the S-carvone concentrations of the tubers from the bottom of the stores were higher than from the top.

12.3.2.2. Processing Quality and Sensory Analyses

The processing quality of different potato cultivars stored in bulk and treated with various Talent applications were analysed during several storage seasons and compared with (C)IPC treated potatoes. No differences in French fry (cv's Bintje and Agria) (Hartmans *et al.* 1995) or chips (cv. Saturna) quality were detected within all these years.

When S-carvone is used as a sprout suppressant for potatoes, it is of importance to know whether this has any influence on the sensory quality ("off-flavour") of steam-cooked or processed potatoes. Several trials carried out during or at the end of the storage season over the last 5 years showed that, if 'Talent' was applied, no changes in taste or flavour occurred of steam-cooked or processed (French fries) potatoes (Hartmans *et al.* 1995).

12.4. S-CARVONE AS A SPROUT GROWTH REGULATOR FOR SEED POTATOES

For longterm storage seed potatoes are in general stored at low temperatures. Cooling is carried out with outside air or by means of the much more expensive mechanical cooling. The minimum storage temperature should be well above 2°C to prevent low temperature injury. In spite of cooling, too excessive sprouting still occurs regularly during storage, dependent of the cultivar and/or storage temperature and storage period.

The sprout suppressing effect of 'Talent' is reversible (Oosterhaven *et al.* 1995d) and can for that reason also be used as a sprout growth regulator of seed potatoes during storage. Various 'Talent' dosages were tested during storage on several cultivars and the effects on sprout suppression, sprout regrowth (sprouting capacity), subsequent crop growth, tuber yield and tuber size distribution has been examined.

12.4.1. Sprout Growth Regulation of Seed Potatoes during Long Term Storage

In (semi)-practical trials with several seed potato cultivars stored in boxes, various 'Talent' dosages were reapplied at regular intervals. The total dose per season did not exceed 250 ml Talent $\cdot$ 1000 kg^{-1}, which was much lower than for ware potatoes. The sprout growth regulation of treated tubers was compared with untreated ones. The best results were achieved when 'Talent' was applied every 6 weeks, 3 to 5 times in total. Furthermore the sprout growth regulating effect of 'Talent' depended of the sprouting characteristics of the cultivars (early or maincrop cultivars, Table 2).

When the tubers were aired before planting, regrowth of the sprouts started. When the sprouting capacity of the seed tubers (Hartmans and van Loon 1987) was determined at that time, no differences has been found between treated or untreated tubers (Table 2).

'Talent' can suppress the growth of already existing sprouts, but depending on the S-carvone concentration they can become dry and necrotic. The apical bud is the least dormant and if the apical bud grows actively it suppresses the growth of the other buds. If the apical bud is becoming necrotic due to 'Talent' treatments, or when it is removed, the other buds start to grow. More sprouts per tuber at planting time can lead

Table 2 Sprout growth regulation of seed potatoes treated between September and April with 'Talent'. (a) sprout growth at unloading time (b) sprouting capacity (sprout weight after 4 weeks at 18°C in darkness) at unloading time

| | Sprout weight (g $\cdot$ tuber^{-1}) | | | |
| | Sprout growth (a) | | Sprouting capacity (b) | |
Cultivar	Control	'Talent'	Control	'Talent'
Early cultivars (*)	3.1	0.3	2.2	2.2
Maincrop cultivars (*)	0.6	0	2.4	2.4

(*) Average values of 6 cultivars.

to more stems per plant which can have an effect on the tuber size distribution at harvest. This regular occurring effect of 'Talent' on seed tubers is comparable with the effect of desprouting, a technique used in practice in order to be able to manipulate the number of sprouts per tuber.

12.4.2. Effect of Carvone Treatment during Storage on Subsequent Crop Growth and Tuber Yield

An effective 'Talent' treatment should reduce sprout growth during storage but should not interfere with subsequent crop growth. Therefore, after planting, emergence, foliage growth, tuber yield and tuber size distribution of treated seed potatoes were studied during several seasons and compared with untreated seed potatoes.

12.4.2.1. Crop Growth

Regular 'Talent' treatments of early sprouting potatoes in general caused a minor delay of a few days on average in 100% emergence. By comparing, after emergence, ground coverage by green foliage, it was shown that 'Talent' treated early seed potatoes sometimes started with a delay in plant development, but afterwards this was compensated with a delay in plant senescence. 'Talent' treated maincrop cultivars during storage showed in general, after planting, no difference in emergence and plant development.

12.4.2.2. Tuber Yield

'Talent' treated seed tubers had, with both early and maincrop potatoes, similar total yields as nontreated tubers (Table 3). Early or midearly cultivars, showed a significant shift in tuber size distribution and had more smaller and less larger tubers after 'Talent' treatments (Table 3). The most important Dutch cultivar Bintje (a midearly cultivar) showed for instance, when treated with 'Talent', in one trial a 20% increase in smaller tubers ($\leqslant 50\,mm$) and a 17% decrease in large tubers ($\geqslant 70\,mm$). This specific effect of 'Talent' treatments during storage appeared to be dependent on: (a) the cultivar used, (b) the time of application and, (c) the amount applied.

Table 3   Influence of 'Talent' treatments during storage of seed potatoes on subsequent tuber yield and tuber size distribution after planting

	Yield kg · plant^{-1}			
	Early cultivars (*)		*Maincrop cultivars* (*)	
Tuber size	*Control*	*'Talent'*	*Control*	*'Talent'*
Total (all)	1.4	1.4	1.36	1.4
$\leqslant 40\,mm$	0.05	0.06	0.06	0.05
40–50 mm	0.16	0.18	0.17	0.18
50–70 mm	0.88	0.91	0.89	0.89
$\geqslant 70\,mm$	0.31	0.25	0.24	0.28

(*) Average yields of 6 cultivars.

It can be concluded that 'Talent', with S-carvone derived from caraway seed as the active ingredient, has good possibilities to be used during storage: (1) as a sprout suppressant for ware potatoes when using higher dosages and (2) as a sprout growth regulator for seed potatoes when using lower dosages.

12.5. BIOCONVERSION OF S-CARVONE AND ITS EFFECT ON WOUND HEALING

12.5.1. Bioconversion of S-carvone

The accumulation of S-carvone in potato tissue of S-carvone treated tubers is dependent on the type of tissue. The maximum residual content of intact tubers in general not exceeds $5 \, mg \cdot kg^{-1}$ fr. wt, as stated earlier. About 90% is associated with the peel fraction, indicating that the peel is a very efficient barrier to S-carvone. The peeled potato tuber fractions contained maximally $0.5 \, mg \cdot kg^{-1}$ fr. wt, and bioconversion products were not detected inside the tuber. Following exposure to S-carvone, wounded potato tubers (half tubers) accumulated less then $20 \, mg \cdot kg^{-1}$ fr. wt (at 15°C), even after 3 weeks of exposure, but since S-carvone was converted into other more reduced compounds, the total amount of monoterpenes was higher, reaching levels of $50–70 \, mg \cdot kg^{-1}$ fr. wt (Table 4). Sprouts contained about the same concentration of S-carvone and its derivatives as the wounded potato tubers (Table 4).

S-carvone is converted by potato sprout and tuber tissue into more reduced compounds (Figure 5). The conversion is stereoselective: R-carvone is mainly converted into neodihydrocarveol, whereas S-carvone is mainly converted into neoisodihydrocarveol. Minor chloroform-soluble products that are formed during the bioconversion are neodihydrocarveol, isodihydrocarveol and dihydrocarveol isomers, hydroxylated

Table 4 Residual content of S-carvone in different potato tuber tissues after exposure to S-carvone (headspace concentration, $5–10 \, \mu g \, l^{-1}$) for 5 days at 15°C in the dark

Tissue	S-Carvone concentration ($mg \cdot kg^{-1}$ fr. wt)	Conversion products ($mg \cdot kg^{-1}$ fr. wt)
Intact tuber[1]	5	not detected
Peeled tuber[1]	<1	not detected
Peel[1]	10	not detected
Half tuber[2]	20	30
Sprouts[3]	20	30

1. exposure of intact tubers to S-carvone, whereafter the three samples were analysed separately.
2. exposure of wounded (half) tubers to S-carvone.
3. exposure of sprouts, in the one-eye model system, to S-carvone.

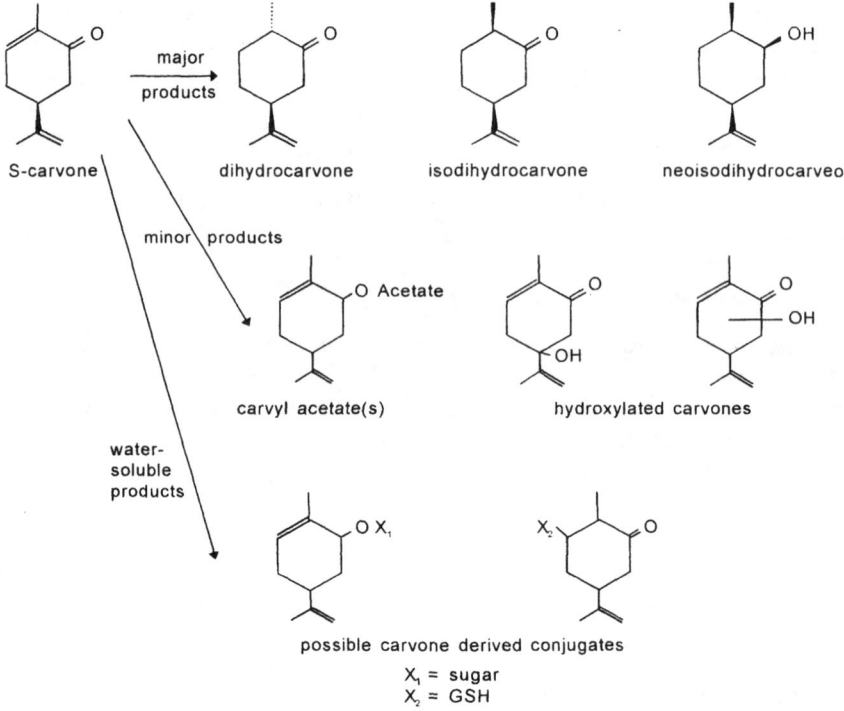

Figure 5   Bioconversion products of S-carvone found in potato tuber tissue

carvones and carvyl acetates. Figure 5 summarizes the compounds detected in potato sprouts.

S-carvone applied to whole tubers is mainly adsorbed at the skin where it is not converted, as bioconversion takes place only in metabolically active tissue. This implies that bioconversion is not playing an important role in the practical use of S-carvone. Applied as sprout growth regulator, S-carvone will be converted rapidly by the already existing sprouts. Thus an increased or reapplied dosage of S-carvone is needed for effective inhibition, but on the other hand it provides the advantage of a reversible suppressant effect on sprout growth.

12.5.2. S-carvone Interferes with the Mevalonate Pathway

Exposure of potato sprouts to S-carvone vapour leads to an inhibition of the longitudinal sprout growth (Oosterhaven *et al.* 1993). The inhibition of elongation of potato sprouts following S-carvone treatment correlates strongly with the decreasing activity of 3-hydroxy-3-methylglutaryl coenzyme A reductase (HMGR; E.C. 1.1.1.34) (Figure 6), a key enzyme in the mevalonate pathway. This effect seems rather specific since enzymes of the citric acid cycle, such as isocitrate dehydrogenase and malate dehydrogenase, were not affected whereas the activity of glutathione related enzymes was stimulated (Oosterhaven *et al.* 1993, Oosterhaven 1995).

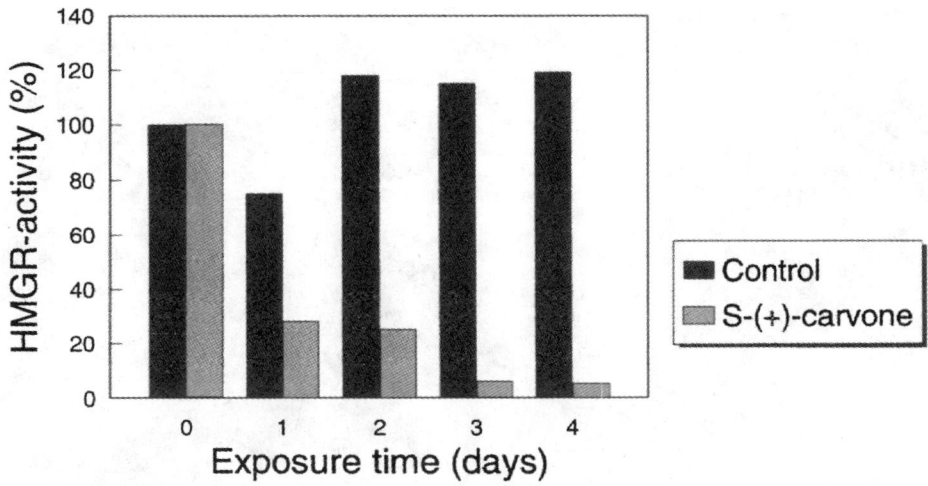

Figure 6 The specific activity of HMGR in organelle fractions following treatment of potato sprouts with S-carvone. The activity is expressed as the percentage of enzyme activity of the control at day zero. A 100% activity corresponds with 30.0 nmol mevalonate $h^{-1} \cdot mg^{-1}$ protein

The mevalonate pathway provides several important compounds, such as hormones, vitamins, carotenoids, membrane components etc., necessary for growth and development (Bach 1987). The (lack of) HMGR activity may already explain the reduced sprout growth. Although the mRNA level remained at a high level, the presence of the HMGR protein could not be detected, using Western blotting techniques, in extracts from sprout tissue exposed to S-carvone for 3 days (Oosterhaven *et al.* 1993). This implies that the synthesis of HMGR is blocked or that the protein is rapidly degraded. However, protein synthesis in general is not blocked, since *in vitro* experiments have shown that translation as a process is not inhibited by S-carvone (Oosterhaven *et al.* 1993). Furthermore, the activity of several enzymes is induced in tuber tissue following an S-carvone treatment, probably together with an increased synthesis of these enzymes, e.g. glutathione reductase, glutathione-S-transferase, dehydroascorbate reductase, S-carvone converting enzymes etc. However, the exact nature of the interference of S-carvone with HMGR remains to be determined.

12.5.3. Interference of S-carvone with the Potato Wound Healing Process

The curing or wound healing period immediately after harvest of potatoes is the period in which the wounds, that occur during harvest, are healed. This is an essential phase of potato storage since wounds are ideal entries for pathogens and because water can easily evaporate from the wounded tubers. S-carvone inhibits the process of wound healing temporarily (Figure 7a and b) by reduction of the suberization and the cambium layer formation. Suberization as a process is not impossible in the presence of S-carvone, since S-carvone-containing tissue develops clearly visible suberin layers after 10–14 days, i.e. after a delay of about 10 days. Initially, S-carvone inhibits the induction

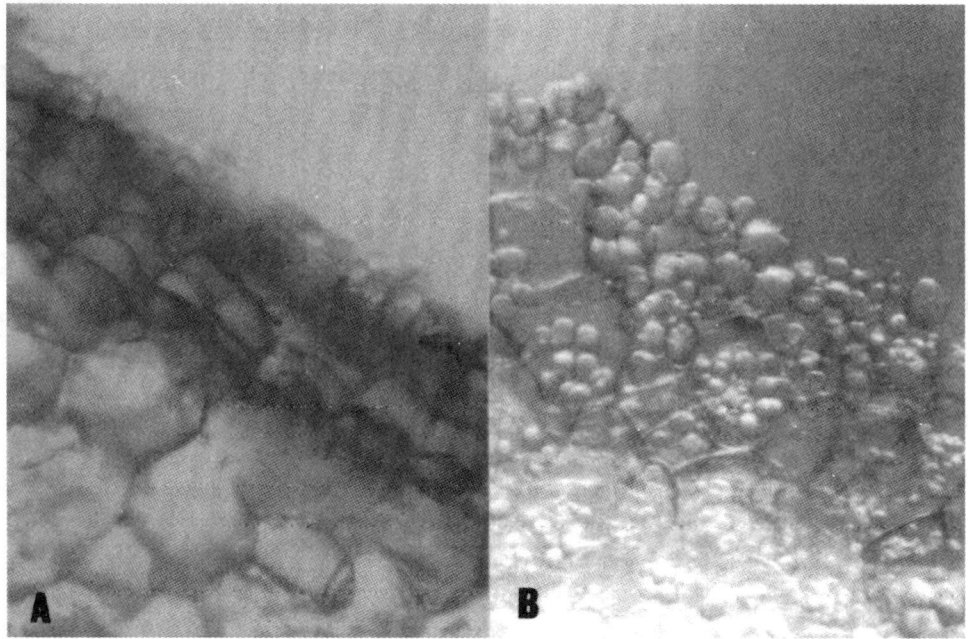

Figure 7   Wounded tissue of potato tuber after 24 days of wound healing. Control tissue (A) and tissue exposed to S-carvone (B). The suberin cell layer was stained with Sudan III (See Color Plate III)

of phenylalanine ammonialyase (PAL) (Oosterhaven *et al.* 1995c) but after about 10 days, the PAL activity increases and suberin is formed.

The inhibition of wound healing is coincided with a lack of induction of 3-hydroxy-3-methylglutaryl coenzyme A reductase (HMGR). Untreated wounded tissue showed an induction of HMGR whereas in S-carvone treated tissue no HMGR could be determined (Oosterhaven *et al.* 1995a). Western-blotting experiments revealed that this was caused by the fact that no HMGR-protein is synthesized in S-carvone containing tuber tissue (Oosterhaven 1995).

The inhibiting effect of S-carvone on wound healing implies that the application of S-carvone-containing sprout suppressants should not be performed at a high dosage before the end of the curing period, because this would lead to a delay of wound healing. In practical trials, a low dosage immediately after harvest has not shown a negative effect on wound healing, nor on pathogen attack or on weight losses (Hartmans *et al.* 1995).

12.6. ANTIFUNGAL ACTIVITY OF S-CARVONE

12.6.1. Mechanism of Antifungal Action of S-carvone

The antimicrobial activity of several essential oils and isolated compounds thereof, such as S-carvone, has been established for many different taxonomic groups of microorganism ranging from Gram-positive and Gram-negative bacteria to fungi including yeasts. Because essential oils contain a variety of compounds from different chemical classes, it is not possible to discriminate one single type of mechanism by which these compounds act on microorganisms. A prominent feature these compounds have in common, is their high degree of hydrophobicity. Due to this feature, these compounds partition preferentially into biological lipid bilayers as a function of their own lipophilicity and the fluidity of the membrane (Oosterhaven 1995b). Accumulation of lipophilic compounds into biological membranes enhances their availability to the cell and thus may cause toxic effects (Sikkema *et al.* 1992, Sikkema *et al.* 1994). This is exemplified by lipophilic hydrocarbons such as ß-pinene and cyclohexane which have been shown to impair energy transducing processes in plasma and mitochondrial membranes of yeast cells (Uribe *et al.* 1985, Uribe *et al.* 1990). Inhibition of yeast mitochondrial respiration by ß-pinene is most likely to be exerted on the level of NADH- and succinate oxidation (Uribe *et al.* 1985).

Despite the high degree of ordering of solutes in the lipid bilayer as compared to the bulk liquid phase (Simon *et al.* 1979), a good correlation has been observed between the partitioning coefficient of various lipophilic compounds in membrane/buffer and octanol/water two-phase systems (Sikkema *et al.* 1994, Sikkema *et al.* 1995). Therefore, octanol/water partitioning coefficients, which are known for many different compounds present in essential oils, can be used to assess the potential antimicrobial activity of these compounds (Sikkema *et al.* 1994). However, the presence of specific reactive groups in such compounds, the variability in membrane composition and the differences in metabolic capacities of target organisms makes a reliable prediction of the antimicrobial activity of these compounds difficult, if not impossible when solely based on their hydrophobicity. Illustrative for this is the action of S-carvone and *trans*-cinnamaldehyde, two compounds with comparable hydrophobicity but a quite different antimicrobial mechanism. Both compounds inhibited *in vitro* growth of the fungus *Penicillium hirsutum* when administered via the gas phase (Smid *et al.* 1995). S-carvone caused full suppression of fungal growth only as long as the compound was present in the gas phase. Cinnamaldehyde, in contrast, caused irreversible inhibition of fungal growth, even after short term exposure. Comparable effects were observed when cell suspensions of the yeast *Saccharomyces cerevisiae* were exposed to different concentrations of S-carvone and *trans*-cinnamaldehyde (Smid *et al.* 1996). Exposure of washed yeast cells for 1 hour to 2 mM *trans*-cinnamaldehyde resulted in 95% loss of viability, whereas S-carvone up to concentrations of 10 mM did not significantly reduce viability of *S. cerevisiae* cells. Evidently, S-carvone acts predominantly as a fungistatic agent, whereas cinnamaldehyde is a fungicidal agent.

This difference between S-carvone and *trans*-cinnamaldehyde in mechanism of antifungal activity was studied in more detail using *Saccharomyces cerevisiae* (Smid *et al.* 1996). Cinnamaldehyde was found to cause a (partial) collapse of the integrity of the cytoplasmic membrane that eventually leads to an excessive leakage of metabolites and enzymes from the affected cells and finally loss of viability. Loss of membrane integrity was not observed with S-carvone which is in agreement with its fungistatic rather than fungicidal effect (Smid *et al.* 1996).

12.6.2. Inhibitory Action of S-carvone towards Plant Pathogenic Fungi

In vitro studies have shown that S-carvone is active against a broad range of plant pathogenic fungi, including *Penicillium hirsutum*, *P. allii* (Smid *et al.* 1994, Smid *et al.* 1995), *Aspergillus parasiticus* (Farag *et al.* 1989), *Botrytis cinerea* and *Rhizopus stolonifer* (Caccioni and Guizzardi, 1994), as well as *Fusarium sulphureum*, *Phoma exigua* and *Helminthosporium solani* (Gorris and Smid, 1996). This broad activity spectrum and the low mammalian toxicity (Jenner *et al.* 1964) establish the potential of this compound to be developed as a natural crop protectant.

Despite the broad activity range of S-carvone, a notable variation in sensitivity towards this compound was observed for different fungal species. Full suppression of germination of *P. hirsutum* conidia was observed in the presence of 40–50 mg/L S-carvone for at least 4 weeks. In this period, the vitality of the spores did not significantly change. Moreover, instant germination and subsequent formation of fungal mycelium was observed after transfer of the spores to a fresh agar medium in a petridish without S-carvone (Smid *et al.* 1995). With *H. solani*, exposed to S-carvone (±55 ppm, for 45 hours) a complete loss of viability of the fungal conidia was noted (Figure 8).

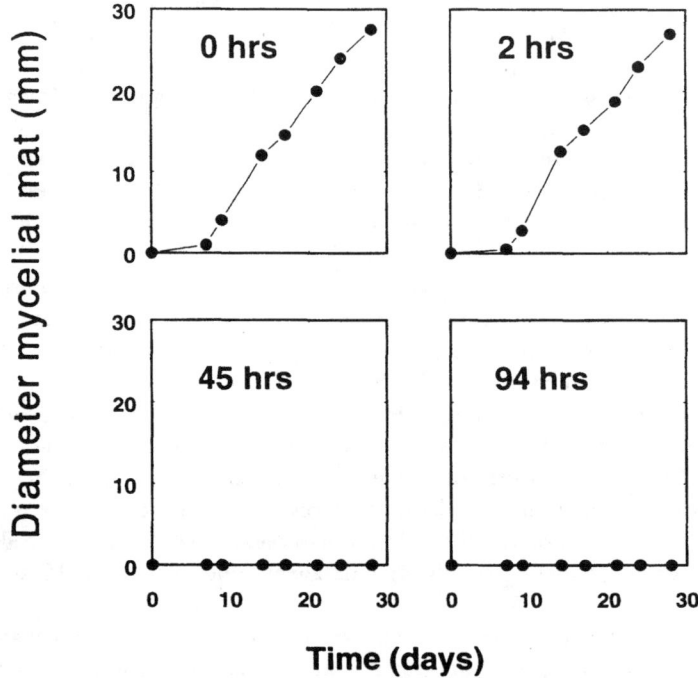

Figure 8 The effect of exposure to S-carvone on radial growth of *H. solani* on potato dextrose agar (PDA) plates at 20°C. Conidia were exposed to S-carvone (45–69 μg/l) for 0 hrs (upper left panel), 2 hrs (upper right panel), 45 hrs (lower left panel) and 94 hrs (lower right panel). After exposure, conidia were transferred to fresh potato dextrose agar plates and incubated at 20°C in the absence of S-carvone

These results show that some fungi, such as *H. solani*, are extremely sensitive towards S-carvone whereas others, like *P. hirsutum* are not. Notably, also fungi with a high degree of tolerance towards the inhibitory effect of S-carvone have been identified. *Fusarium solani* var. *coeruleum* is, in contrast to the closely related *F. sulphureum*, relatively insensitive towards S-carvone (Gorris *et al.* 1994). This tolerance was particularly apparent when the fungus was incubated *in situ* on potato tubers (Gorris and Smid 1996). In a follow-up study it was shown that both *Fusarium* species converted S-carvone at comparable rates into, mainly isodihydrocarvone, isodihydrocarveol and neoisodihydrocarveol (Oosterhaven *et al.* 1996). Since the individual conversion products do not show antifungal activity, chemical reduction of S-carvone can be considered as a detoxification process for the fungus. The bioconversion of S-carvone as executed by *F. solani* var. *coeruleum* and *F. sulphureum* does not offer an explanation for the observed difference in sensitivity towards S-carvone.

12.6.3. S-carvone as Postharvest Crop Protecting Agent

S-carvone has been tested on laboratory scale for its *in situ* efficacy to control postharvest diseases of tulip bulbs (Smid *et al.* 1995) and potato tubers (Gorris *et al.* 1994). In these studies S-carvone was allowed to vaporize in the head space of hermetically closed test containers to concentrations of approximately 50 mg/L. It was shown that under these conditions treatment with S-carvone for one week completely suppressed *Penicillium* rot on tulip bulbs, while 68% of the non-treated control bulbs were infected. Conceivable, the phytotoxic effects of S-carvone on flower development have been tested as part of the research. No apparent phytotoxic effects were found that would adversely affect the marketing quality of the tulip bulbs (Smid *et al.* 1995).

The suppressive effect of S-carvone, administered via the gas phase, on disease development of stored potato tubers has been demonstrated for several phytopathogenic fungi (Hartmans *et al.* 1995). In closed containers and at 9°C, an S-carvone concentration of approximately 35 mg/L could be maintained for almost 4 weeks (Figure 9 insert). After inoculation of the artificially wounded tubers with *P. exigua* spores, the tubers were stored for 7 weeks at 9°C in either the absence or presence of S-carvone. Storage in the presence of S-carvone resulted in a considerable reduction of the number of infected tubers at each inoculum dose tested (Figure 9). However, at a low inoculum dose (from 2.5 to 10 spores applied per wound site), relatively the highest reduction in disease development was observed. Overall, a considerable suppression of lesion development was observed with tubers stored in the presence of S-carvone (Slotboom and Smid, unpublished). Comparable results were obtained with tubers artificially infected with *F. sulphureum*. In contrast, under the same experimental conditions, no effect of S-carvone on disease development was observed with tubers inoculated with *F. solani* var. *coeruleum*.

Silver scurf considered to be a typical postharvest disease of potatoes and is caused by the *Helminthosporoum solani* fungus. The fungus actually penetrates the periderm of the tuber and under optimal conditions, hyphae can be found in the phellem, phelloderm and cortex of infected tubers (Jellis and Taylor 1977). During prolonged storage, excessive fresh weight losses may result from *H. solani* infected tubers (Jellis and Taylor

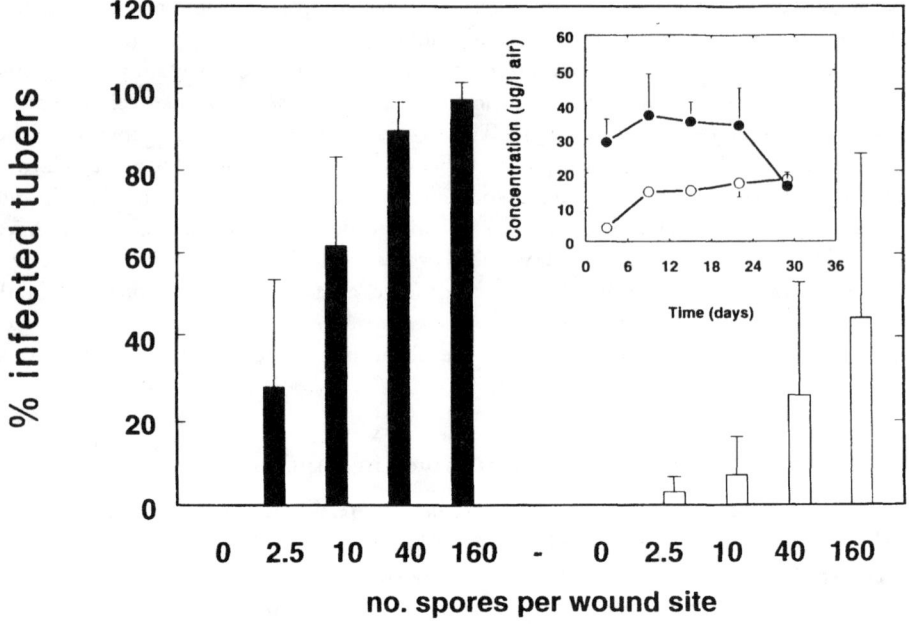

Figure 9 Suppression of *Phoma exigua* on potato tubers (cultivar Bintje, size 40–45) by S-carvone. Tubers were artificially wounded and wound sites were inoculated with 0, 2.5, 10, 40 and 160 *P. exigua* spores. Next, the tubers were stored for 35 days at 9°C in closed containers in the absence (closed bars) and presence (open bars) of S-carvone. During storage, the S-carvone (insert, closed symbols) and dihydrocarvone (insert; open symbols) concentration was monitored using GC-analysis of the head space. The standard error of the mean is indicated by error bars. (Slotboom and Smid, unpublished)

1977). Storage experiments were performed in our laboratory to demonstrate the *in situ* activity of S-carvone to suppress the development of silver scurf on potato tubers. Naturally infected tubers were stored at 9°C and 18°C, under high relative humidity (97–99%) in the presence and absence of S-carvone. At 9°C, S-carvone completely blocked the formation of conidiophores and conidia on the surface of tubers (Slotboom and Smid, unpublished). At 18°C, reduction of approximately 50% of the formation of conidia was observed. In storage trials on semi-practical scale, S-carvone treatment was found to suppress silver scurf for up to 90% compared to untreated controls.

12.7. FUTURE OUTLOOK

The farmers from the high Andes showed us that their traditional knowledge of allelopathic characteristics of certain plants was interesting and could also be of great importance for the development of other new natural crop protection products, as it did for the development of S-carvone as a sprout growth suppressant or regulator for (seed)potatoes.

S-carvone is a natural compound and is approved for application in food as a flavouring agent. The compound is by placed in catagory A. by the Council of Europe (C.E.), indicating that application of the compound in food is permitted to levels not exceeding concentrations which are needed for the intended use. In addition, S-carvone, as a flavouring agent, is also present on the so-called GRAS (Generally Recognised as Safe) list published by FEMA (Flavor and Extract Munufacturers' Association) and FDA (Food and Drug Administration, USA). This paper shows that the effective use in practice of S-carvone as an antifungal agent for postharvest crop protection has been established in a number of cases. Whenever a plant is considered to be exploited as a green chemical source, a thorough evaluation will have to be carried out of its value with respect to the net economy of its cultivation and actual production of the green chemical (be it the whole crop, an extract or a purified compound), market value of the antimicrobial preparation and costs for going through all obligatory legislative procedures. Not many of the potential sources may then pass this evaluation. The economy of changing from the range of available synthetic chemicals to green chemicals will in part determine its market value and will ultimately dictate whether commercialization is feasible. The recent increase in the legislative pressure towards non-chemicals may favour these economic odds.

REFERENCES

Aliaga, T.J. and Feldheim, W. (1985). Hemmung der Keimbildung bei gelagerte Kartoffen durch das ätherische Öl der südamerikanischen Muñapflanze (*Minthostachis* spp.). *Ernährung/Nutrition*, **9**, 254–256.

Ausher, R. (1996). Integrated Pest Management. *Outlook on Agriculture*, **25**, 107–113.

Bach, T.J. (1987). Synthesis and metabolism of mevalonic acid in plants. *Plant Physiology and Biochemistry*, **25**, 163–178.

Baerheim Svendsen, A., Scheffer, J.J.C. and Looman, A. (1987). Composition of the Volatile Oil of *Minthostachys glabrescens* Epl. *Flavour and Fragrance Journal*, **2**, 45–46.

Beveridge, J.L., Dalziel, J. and Duncan, H.J. (1981). The assessment of volatile organic compounds as sprout suppressant for ware and seed potatoes. *Potato Research*, **24**, 61–76.

Bournot, K. (1949). Über die Einwirkung von Duftstoffen auf den lebenden Organismus, insbesondere den der Pflanzen. *Pharmazie Zeitschrift-Berlin*, **4**, 81–86.

Boyd, W.D. and Duncan, H.J. (1986). Studies on potato sprout suppressants. 7. Headspace and residue analysis of chlorpropham in a commercial box potato store. *Potato Research*, **29**, 217–223.

Burton, W.G. (1952). Studies on the dormancy and sprouting of potatoes. III The effect upon sprouting of volatile metabolic products other than carbon dioxide. *New Phytology*, **51**, 154–161.

Burton, W.G. (1955). Biological and economic aspects of the refrigerated storage of potatoes. *Proceedings of the Institute of Refrigeration*, **51**, 168–172.

Burton, W.G. (1965a). The sugar balance in some British potato varieties during storage. I Preliminary observations. *European Potato Journal*, **8**, 80–91.

Burton, W.G. (1965b). The effect of oxygen and the volatile products of metabolism upon the sprout growth of potatoes. *European Potato Journal*, **8**, 245.

Burton, W.G. and Meigh, D.F. (1971). The production of growth-suppressing volatile substances by stored potato tubers. *Potato Research*, **14**, 96–101.

Caccioni, D.R.L. and Guizzardi, M. (1994) Inhibition of germination and growth of fruit and vegetable postharvest pathogenic fungi by essential oil components. *Journal of Essential Oil Research*, **6**, 173–179.

Chan, P.C. (1990). NTP technical report on the toxicology and carcinogenesis studies of d-carvone (CAS) No. 2244-16-8) in B6C3F1 mice. NIH publication No. 90-2836, US Department of Health and Human Services.

Farag, R.S., Daw, Z.Y. and Abo Raya, S.H. (1989) Influence of some spice essential oils on *Aspergillus parasiticus* growth and production of aflatoxins in a synthetic medium. *Journal of Food Science*, **54**, 74–76.

Friedman, J. (1987). Allelopathy in Desert Ecosystems. In: *Allelochemicals: Role in Agriculture and Forestery*. Ed. Waller, G. R.; American Chemical Society Washington D.C.

Gorris, L.G.M., Oosterhaven, K., Hartmans, K.J., De Witte, Y. and Smid, E.J. (1994) Control of fungal storage diseases of potato by use of plant-essential oil components. *Brighton Crop Protection Conference, Pests and Diseases*, 307–312.

Gorris, L.G.M. and Smid, E.J. (1996) The use of natural compounds for crop protection. In: Crop Protection in Northern Britain Vol. I, SCRI, Dundee, UK, 133–138.

Hartmans, K.J. and Van Es A. (1988). Alternative sprout inhibitors. *Proceedings of the 22nd annual conference of the European Chips and Snacks Association*- PRC, Heelsum. The Netherlands, pp. 102–103.

Hartmans, K.J. and Van Loon C.D. (1987). Effect of physiological age on growth vigour of seed potatoes of two cultivars. 1. Influence of storage period and temperature on sprouting characteristics. *Potato Research*, **30**, 397–409.

Hartmans, K.J., Diepenhorst, P., Bakker, W. and Gorris. L.G.M. (1995). The use of carvone in agriculture: sprout suppression of potatoes and antifungal activity against potato tuber and other plant diseases. *Industrial Crops and Products*, **4**, 3–13.

Heath, H.B. (1973). *Flavor Technology: Profiles, Products, Applications.* AVI Publishing Company, Inc., Westport, Connecticut. pp. 122–124.

Janssen, A.M. (1989). Antimicrobial activities of essential oils – a pharmacognostical study. Ph.D. Thesis, Leiden University, pp. 141–162.

Jellis, G.J. and Taylor, G.S. (1977) The development of silver scurf (*Helminthosporium solani*) disease of potato. *Annals of Applied Biology*, **86**, 19–28.

Jenner, P.M., Hagan, E.C., Taylor, J.M., Cook, E.L. and Fitzhugh, O.G. (1964) Food flavourings and compounds of related structure. I. Acute oral toxicity. *Food and Cosmetic Toxicology*, **2**, 327–343.

Marth, P.C. and Schultz, E.S. (1952). A new sprout inhibitor for potato tubers. *American Potato Journal*, **29**, 268–278.

Meigh, D.F. (1969). Suppression of sprouting in stored potatoes by volatile organic compounds. *Journal of the Science of Food and Agriculture*, **20**, 159–164.

Meigh, D.F., Filmer, A.A.E and Self, R. (1973). Growth inhibitory volatile aromatic compounds produced by *Solanum tuberosum* tubers. *Phytochemistry*, **12**, 987–993.

Molisch, H. (1937). *Der einfluss einer Pflanze auf die andere – Allelopathy*; Gustaf Fischer: Jena.

Oosterhaven, J. (1995) Different aspects of S-carvone; A natural potato sprout growth inhibitor. Thesis Wageningen, ISBN 90-5485-435-9.

Oosterhaven, K., Hartmans, K.J. and Huizing, H.J. (1993). Inhibition of potato (*Solanum tuberosum*) sprout growth by the monoterpene S-carvone: reduction of 3-hydroxy-3-methylglutaryl coenzyme A reductase activity without effect on its mRNA level. *Journal of Plant Physiology*, **141**, 463–469.

Oosterhaven, K., Hartmans, K.J., Scheffer, J.J.C. and Van der Plas, L.H.W. (1995a). The monoterpene S-carvone inhibits wound healing of wounded potato tuber tissue. *Physiologia Plantarum*, **93**, 225–232.

Oosterhaven, K., Poolman, B. and Smid, E.J. (1995b). S-carvone as a natural fungistatic, bacteristatic and potato sprout inhibiting compound. *Industrial Crops and Products*, **4**, 23–31.

Oosterhaven, K., Hartmans, K.J., Scheffer, J.J.C. and Van der Plas, L.H.W. (1995c). S-carvone inhibits phenylalanine ammonia lyase (PAL) activity and suberization during wound healing of potato tubers. *Journal of Plant Physiology*, **146**, 288–294.

Oosterhaven, K., Hartmans, K.J. and Scheffer, J.J.C. (1995d). Inhibition of potato sprout growth by carvone enantiomers and their bioconversion in sprouts. *Potato Research*, **38**, 211–222.

Oosterhaven, K., Chambel Leitao, A., Gorris, L.G.M. and Smid, E.J. (1996) Comparative study on the action of S-(+)-carvone, *in situ*, on the potato storage fungi *Fusarium solani* var. *coeruleum* and *F. sulphureum*. *Journal of Applied Bacteriology*, **80**, 535–539.

Ormachea, E.C.A. (1979). Usos tradicionales de la "muña" (Minthostachys spp., Labiatae) en aspectos fitosanitarios de Cusco y Puno. *Revista Peruana de Entomologia*, **22**, 67–69.

Rastovksi, A. (1987). Ventilation systems in potato stores. In *Storage of Potatoes*. Ed. Rastovski, A., van Es, A., *et al.*; Pudoc Wageningen, The Netherlands, pp. 286–309.

Sikkema, J., Poolman, B., Konings, W.N. and De Bont, J.A.M. (1992) Effects of the membrane action of tetralin on the functional and structural properties of artificial and bacterial membranes. *Journal of Bacteriology*, **174**, 2986–2992.

Sikkema, J., De Bont, J.A.M. and Poolman, B. (1994) Interactions of cyclic hydrocarbons with biological membranes. *Journal of Biological Chemistry*, **269**, 8022–8028.

Sikkema, J., De Bont, J.A.M. and Poolman, B. (1995) Mechanisms of membrane toxicity of hydrocarbons. *Microbiol. Review*, **59**, 201–222.

Simon, S.A., Stone, W.L. and Bennett, P.B. (1979) Can regular solution theory be applied to lipid bilayer membranes? *Biochim. Biophys. Acta*, **550**, 38–47.

Smid, E.J., De Witte, Y., Vrees, O. and Gorris, L.G.M. (1994) Use of secondary plant metabolites for the control of postharvest fungal diseases on flower bulbs. *Acta Hort.*, **368**, 523–530.

Smid, E.J., De Witte, Y. and Gorris, L.G.M. (1995) Secondary plant metabolites as control agents of postharvest *Penicillium* rot on tulip bulbs. *Postharvest Biol. Technol.*, **6**, 303–312.

Smid, E.J., Koeken, J.G.P. and Gorris, L.G.M. (1996) Fungicidal and fungistatic action of the secondary plant metabolites cinnamaldehyde and carvone. In H. Lyr, P.E. Russell and H.D. Sisler, (eds.), *Modern Fungicides and Antifungal Compounds*, Intercept Ltd., Andover, U.K. pp. 173–180.

Stoll, G. (1986). *Natural crop protection in the tropics*. Josef Margraf Publishers, Aichtal Germany, pp 148–151.

Uribe, S., Ramirez, J. and Pena, A. (1985) Effects of β-pinene on yeast membrane functions. *Journal of Bacteriology*, **161**, 1195–1200.

Uribe, S., Rangel, P., Espinola, G. and Aguirre, G. (1990) Effects of cyclohexane, an industrial solvent, on the yeast *Saccharomyces cerevisiae* and on isolated yeast mitochondria. *Applied Environmental Microbiology*, **56**, 2114–2119.

Vaughn, S.F. and Spencer, G.F. (1991). Volatile monoterpenes inhibit potato tuber sprouting. *American Potato Journal*, **68**, 821–831.

Vaughn, S.F. and Spencer, G.F. (1993). Naturally-occurring aromatic compounds inhibit potato tuber sprouting. *American Potato Journal*, **70**, 527–533.

Zuelsdorff, N.T. and Burkholder, W.E. (1978). Toxicity and repellency of Umbelliferae plant compounds to the granary weevil, *Sitophilus granarius*. *Proceedings of the North Central Branch of the Entomological Society of America*. 75th Annual Conference of the North Central States Entomologists, **33**. 28.

INDEX

197

Color Plate I. *See* Zenon Wêglarz, Figure 8.2, Page 132.

Color Plate II. *See* Klaasje J. Hartmans *et al.*, Figure 12.3, Page 180.

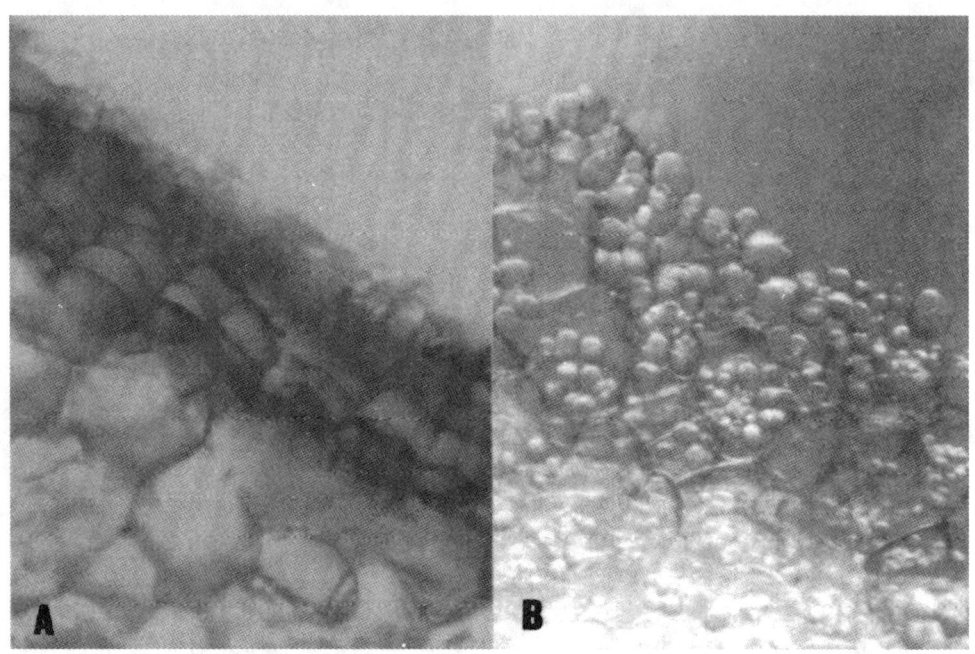

Color Plate III. *See* Klaasje J. Hartmans *et al.*, Figure 12.7, Page 188.